Hellenic Studies 59

THE THEOLOGY OF ARITHMETIC

Recent Titles in the Hellenic Studies Series

THE THEOLOGY OF ARITHMETIC

*Number Symbolism in
Platonism and Early Christianity*

Joel Kalvesmaki

CENTER FOR HELLENIC STUDIES
Trustees for Harvard University
Washington, DC
Distributed by Harvard University Press
Cambridge, Massachusetts, and London, England
2013

The Theology of Arithmetic: Number Symbolism in Platonism and Early
    Christianity
        By Joel Kalvesmaki
Copyright © 2013 Center for Hellenic Studies, Trustees for Harvard University
All Rights Reserved.
Published by Center for Hellenic Studies, Trustees for Harvard University,
    Washington, D.C.
Distributed by Harvard University Press, Cambridge, Massachusetts and
    London, England

Cataloging-in-Publication data available from the Library of Congress.
ISBN: 9780674073302

Contents

Contents

Acknowledgments

O F THOSE WHO HAVE HELPED ME IN MY RESEARCH, the most important is my wife, Colette, who allowed me to work undisturbed, and whose curiosity about my topic provided many stimulating conversations. Robin Darling Young was the first to suggest that the topic would be fertile; her intuition was correct, probably far more than she knew. Early drafts of chapters were read by Kevin Corrigan, Andrew Hill, Christoph Markschies, William McCarthy, Ellen Muehlberger, John Nesbitt, Philip Rousseau, Janet Timbie, John Turner, Susan Wessel, and Michael Williams, who each in his or her own way provided welcome remarks. I appreciate the comments of my (alas) anonymous readers and of the professionals behind the editorial and design phases of the book: Scott Johnson, Jill Curry Robbins, and David M. Weeks. I thank everyone above for their stimulating suggestions and criticism, which have been critical to a narrative that I hope is comprehensible, intellectually engaging, and a pleasure to read.

Abbreviations

ANF	Roberts, A., J. Donaldson, and A. C. Coxe, eds. 1885–1896. *Ante-Nicene Fathers*. Buffalo.
BCNH.É	Bibliothèque copte de Nag Hammadi. Section "Études."
BCNH.T	Bibliothèque copte de Nag Hammadi. Section "Textes."
CAG	Commentaria in Aristotelem Graeca
CSEL	Corpus Scriptorum Ecclesiasticorum Latinorum
DECL	Döpp, S., and W. Geerlings, eds. 2000. *Dictionary of Early Christian Literature*. Trans. M. O'Connell. New York.
FOTC	Fathers of the Church
GCS	Die Griechischen Christlichen Schriftsteller
IGLS	*Inscriptions grecques et latines de la Syrie*
KRS	Kirk et al. 1983
LSJ	Liddell, H. G., and R. Scott. 1996. *A Greek-English Lexicon*. Rev. H. S. Jones et al. 9th ed. Oxford.
LCL	Loeb Classical Library
LXX	Septuagint
NH	Nag Hammadi
NHS	Nag Hammadi [and Manichaean] Studies
OCD	Hornblower, S., and A. Spawforth, eds. 1996. *Oxford Classical Dictionary*. 3rd ed. Oxford.
PG	Patrologiae cursus completus, Series graeca. 161 vols. in 166 pts.
PRE	*Pauly's Real-Encyclopädie der classischen Altertumswissenschaft*. 1837–1852. Stuttgart.
SBL	Society of Biblical Literature
SC	Sources chrétiennes
SVF	von Arnim, H. F., ed. 1968. *Stoicorum veterum fragmenta*. 4 vols. Stuttgart.
TLG	Thesaurus Linguae Graecae
TRE	Krause, G., and G. Müller, eds. 1977–. *Theologische Realenzyklopädie*. Berlin.
WUNT	Wissenschaftliche Untersuchungen zum Neuen Testament

1

Introduction

IN ANCIENT JUDAISM AND CHRISTIANITY, the fascination with numbers and number symbolism was widespread. Just think of the 12 tribes and 12 apostles, Enoch's 365 years on earth, and 40-day periods of fasting or mourning. In such patterns, Jews and Christians shared a common vocabulary with all ancient societies, which used numbers to adorn their lore, order their calendars, and frame their cosmology. In the second century, a variety of new Christian movements—among the most prominent one called the Valentinians—developed systems of number symbolism that went well beyond what is found in Scripture and early Jewish and Christian lore, both in meaning and intensity. These Christians described God and interpreted the Bible through the lens of arithmetic. To them God was a 'unit' or 'monad,' who had self-erupted into a harmonic multiplicity of 'aeons' or 'emanations,' from there showering down upon all levels of reality, both material and immaterial worlds, an arithmetical structure rooted in the divinity. Scriptural verses were used to justify and explain this theology of arithmetic, and the Bible was treated as a repository of numbers, a cache of proof texts that could be understood only in the light of secret revelation.

The principal impetus for this theology of arithmetic was the Platonic and Pythagorean literary tradition, in which numbers were a driving force of metaphysics, symbolism, and interpretation. Philosophers in the century Jesus lived in argued over the number of principles in the universe, and intellectuals such as Plutarch used numbers and their symbolic meanings to interpret religious myths and to frame the natural world. Platonist and Pythagorean number symbolism provided a credible framework for some second- and third-century Christians to build their new theologies. They continued from neo-Pythagorean deliberations—Do things start with a monad or a dyad? If the latter, what kind?—and put forward a conflicting variety of philosophical myths, intellectual archways erected seemingly ad hoc, both to straddle and to intrude upon the structures of Middle Platonism (today so called) and Christianity.

The ecclesiastical dispute that emerged over the appropriate role of numbers was brief but intense. The Valentinians were fiercely opposed by orthodox apologists such as Irenaeus, a second-century bishop of Lyons whose major work, *Against Heresies*, is one of the most important sources for this dispute. Irenaeus directly tackled the Valentinians' use of numbers, and argued that numbers should be subservient to the tradition handed down by the apostles, not vice versa. His argument against the Valentinians was echoed by other Christians at the time, and so Irenaeus' argument and tone articulately express opposition to Valentinian theology across the Mediterranean. But that reaction belies the widespread use of number symbolism by orthodox Christians. The orthodox culture of number symbolism can be seen in the writings of Irenaeus, and especially of his near contemporary Clement of Alexandria, who opposed the Valentinians for similar reasons, but who was inclined to correct by example, to counter Valentinian errors tacitly. Irenaeus and Clement of Alexandria typified the number symbolism that was to circulate in the Church for centuries, without controversy, into the medieval period.

The second-century Christian debate over numbers eventually influenced the very tradition that inspired it. Platonist philosophy after Plotinus increasingly depended upon a number symbolism that took on the colors painted by early Christians. And in the fourth century, Theodore of Asine was to some fellow Platonists as much a heretic in his number symbolism as the Valentinians were to other Christians. Iamblichus was the Platonists' Irenaeus and Clement at once, articulating principles for a reasoned use of number symbolism. Although the influence of Christian theology upon the Platonists is not explicitly discussed in our sources, I believe that it is evident in the parallels. Christians were the first to expand the imaginative metaphysical possibilities of number symbolism and the first to engage in a critical discussion about what transgressed their tradition. Both of these Christian trends expanded the cultural horizons of the Mediterranean, and cleared the way for analogous Platonist developments.

That, in outline form, is the argument I advance in this book, written for anyone interested in intellectual history in late antiquity. An astute reader may wonder where Judaism fits into this story. After a substantive look at Philo (chapter 2) and glancing looks at Aristobulus and Hermippus (chapter 7), I say little more about Jewish authors for whom numbers were important, for two reasons. First, the story I tell centers on the role of numbers in the philosophical–theological tradition, particularly metaphysics. That tradition is not attested in the non-Christian Jewish sources that can be reliably dated to the second to fourth centuries. Second, the practice of psephy (see below), discussed so frequently in this book, was developed in Hebrew and Aramaic literature only after it had matured in Greek. That is, Jewish gematria is largely dependent upon, not the cause of, the gematria discussed in this study. This may seem counterintuitive to many readers, especially those who would think the Bible is the source of gematria. It is not. To explain my rationale adequately requires a separate, cultural history of psephic practices and habits, which is in preparation.

Some of my terminology needs a brief explanation. For instance, I tend to avoid the term 'gnostic,' which has been greatly abused over the last century. The term as we use it today, as scholars have increasingly recognized, is a modern invention, and does not adequately describe the fractious and polymorphic religious landscape evident in both Irenaeus' *Against Heresies* and the Nag Hammadi library.[1] Hence I avoid the label altogether, except when the sources warrant (for instance, when Irenaeus or Clement of Alexandria use it; but even there it describes a specific group, not all the heresies). But I do use the relatively vague term 'gnosticizing,' not to separate groups, but to describe behaviors, as explained in the context of my discussion.

It may seem that by focusing first on Christian groups classified today as gnostic (chapters 3–5), and by setting them in opposition to Irenaeus and his followers (chapters 6 and 7), I uncritically adopt Irenaeus' distinctions between heretical and orthodox and between gnostic and orthodox. This judgment would be hasty. The opposition between Irenaeus and these various groups is not my premise but my thesis. I argue, not assume, that the two have conflicting views of the role of numbers. The point should become apparent in my account of the differences. Furthermore, this book is not about every second-century Christian group considered deviant. A number of heresies popularly labeled 'gnostic' fail to bear consideration, since not all systems of gnosis had an interest in number symbolism. Prominently missing from consideration are Simon Magus, Marcion, Basilides, and "Sethian" texts. This is not to say that they do not use numbers as symbols or literary devices (they do, sometimes profusely), but they do not show a propensity to mathematical theology.

I use the term 'orthodox' provisionally, as do most scholars of late antique Christianity who wisely wish to avoid declaring which party was in the right. But despite the good reasons for qualifying the term, 'orthodox' can be used justifiably in this study. The Valentinians and the like-minded used novel ideas to advance provocative theses; their opponents, who came from disparate parts of the Roman Empire, sought to articulate principles that were as old and as widespread as the Church. Valentinians cherished private, elite audiences; Irenaeus, Clement, Tertullian, and Hippolytus appealed to the Church at large. So I use 'orthodox' of men who appealed to a common, lived experience of the

---

[1]   For more on the artifice of the term see M. Williams 1996.

Church—that is, churches that identified themselves with the other churches scattered about the Mediterranean—but I do not mean to comment on how justifiable that claim was.

To describe Valentinian systems I have used 'protology' (and derivatives).[2] The term is helpful because it points to the arithmetical character of Valentinian theories of how everything started from a πρῶτος, a unitary principle. That nuance is missing in synonyms such as 'metaphysics,' 'philosophy,' and 'theology.' But 'protology' can be misleading, since it implies that the Valentinians were doing something wholly separate from that which occupied their orthodox counterparts, whose ruminations on the relationship between the Father and the Son we would more comfortably call 'theology,' not 'protology.' In reality, both groups were vying for the same conceptual space, trying to present a plausible account of what constitutes the highest order of the universe. Therefore I use 'protology' only when it refers to the emergence of multitude from the original πρῶτος (or δεύτερος, as we shall see).

I also distinguish in this study between 'arithmetic' and 'mathematics.' The former, the study of the properties and operations of discrete numbers (e.g. addition and multiplication), is a proper subset of the latter, which is the study of all the numerical sciences. Mathematics comprises arithmetic, geometry, music, and astronomy—the ancient foursome known in Greek as the τέσσαρες μεθόδοι and later in Latin (via Boethius) as the *quadrivium*. Thus, in antiquity, references to mathematicians (such as in Sextus Empiricus' *Against the Mathematicians*) signify those who handle any of the numerical sciences (and who use these sciences to investigate the world), not merely those skilled in computation. This classical distinction between the two words was current in English as late as the eighteenth century. The conceptual overhaul of the sciences in the age of Newton led to our present usage, where 'arithmetic' and 'mathematics' are often interchangeable. For aspects of this study, however, the ancient distinction is helpful, especially since we often encounter the term μαθήματα, which always implies more than our modern term 'mathematics' does.

I use the term 'numerology' only of number symbolism intended either to reveal or to keep concealed occult knowledge. Think of it as a correlate to 'astrology,' which has similar connotations that distinguish it from 'astronomy.' All ancient numerology is number symbolism, but only a part of ancient number symbolism is numerology. Some ancient authors would object, were they to learn that their number symbolism was today conflated with activities such as predicting the outcome of a marriage or determining the date of a person's death based on the numerical value of his name. Thus I generally prefer the

[2]   Orbe 1976:484n198, reinforced throughout in Thomassen 2006.

neutral term 'number symbolism' unless prognostication is at work, in which case 'numerology' is accurate. As a synonym for 'number symbolism' Delatte (1915) introduced the term 'arithmology,' which he found in the title of a Greek text preserved in an eighteenth-century manuscript. His neologism is anachronistic. In the Athens manuscript he consulted (Bibl. Nat. 65), ἀριθμολογία, a hapax legomenon, describes a late or post-Byzantine collection of numbered lists, and the text has no overt number symbolism. But 'arithmology' breaks up the monotony of 'number symbolism,' and it has made its way into our vocabulary, so I use it occasionally as a synonym.

I use 'psephy,' 'psephic,' and 'isopsephy' (all derived from ψῆφος 'pebble' or 'vote') to describe the ancient habit of reckoning the numerical value of names and words based on the arithmetical values of their constituent letters. This is more commonly known in English as 'gematria,' but the Hebrew term that this word comes from was not coined until probably the sixth or seventh century. For the cultural and chronological scope of this study, the first three terms are more appropriate.

In general, references to the Hebrew Scriptures follow the Septuagint, explicitly marked when a difference in content (not just chapter and verse numeration) is salient. Uncredited translations are my own.

This is the first in-depth exploration of early Christian number symbolism. Certainly, there have been dozens of studies about the individual texts and ideas I discuss, and I signal many of these in the notes and bibliography. But no one, to my knowledge, has tried to describe and explain as a whole the late antique Christian and Platonist debates over number symbolism. Although this is the first word, I hope it is not the last. There is much work to be done on this topic. I have not, for instance, analyzed number symbolism in the Barbeliotes, the Peratae, the *Apocryphon of John*, the so-called Ophites, the *Books of Jeu*, the untitled text from the Bruce codex, or *Pistis Sophia*. Closer to orthodox circles are the *Shepherd of Hermas*, the *Sybilline Oracles*, and various apocalyptic texts, which frequently indulge in number symbolism. And the later Christian tradition of number symbolism is rich and understudied. On the Platonist and Pythagorean side, number symbolism is abundant in numerous texts, including the Nag Hammadi treatise *Marsanes*; the numerous commentaries on Plato from late antiquity; and the namesake of my book, the *Theology of Arithmetic*, a bestiary-like collection of numerical lore dating from late antiquity and before. I hope my study stimulates others to explore ancient number symbolism further.

2

Generating the World of Numbers

Pythagorean and Platonist Number Symbolism
in the First Century

TWO INTELLECTUAL TRADITIONS FROM CLASSICAL ANTIQUITY laid the foundation for the early Christian theology of arithmetic. The first, and most easily identified, was the rich tradition of number symbolism in the ancient Mediterranean. Numbers had been used symbolically from very early times and in many cultures, as attested by cuneiform tablets and Egyptian hieroglyphs. In the Greco-Roman world of the first century, strands of number symbolism drawn from different periods and cultures had coalesced into a single amorphous tradition. For inhabitants of the Roman Empire it provided a rich storehouse of ideas with which they could interpret omens, religious texts, the natural world, and the mathematical sciences.

The second great tradition was metaphysical speculation. The earliest Greek philosophers were noted for asking how many sources or roots—ἀρχαί—there were to the universe. Was it monadic, dyadic, polyarchic, or a synthesis? How many levels of reality, if any, existed above the material world? If the universe started out as a monadic unity, how did the second element or level originate? What kind of metaphysical entity were numbers? These questions were discussed by the earliest Greek philosophers, including Plato, but they were addressed with renewed vigor in the first century BCE, when so-called Middle Platonist philosophers, inspired by Pythagorean speculation, put forward new metaphysical theories inextricably linked to the definition and symbolism of numbers.

In this chapter I discuss these two traditions, which early Christian theology brought together. To explain the culture of number symbolism in the first century—the ways numbers were used in antiquity to interpret literature and the natural world—I turn to the writings of Philo of Alexandria and Plutarch of Chaeronaea. Both men incorporate number symbolism into their arguments, and so show us its importance in persuading ancient readers to adopt a certain perspective or interpretation. In the second half of the chapter I turn to the Greek philosophical tradition. After setting out in very broad strokes the metaphysical issues that concerned classical philosophers, I focus on the innovations of early Middle Platonism, namely the Pythagorean and Platonist philosophies described by Alexander Polyhistor, Eudorus of Alexandria, and Moderatus of Gades. I offer new interpretations of texts written by the last two, and in so doing I explain how Greek metaphysics took a turn toward monism and numerically patterned metaphysics.

Numbers to Interpret the World: Philo of Alexandria and Plutarch of Chaeronea

In the first century, the rich mélange of number symbolism that circulated in the Mediterranean was associated with Pythagoras, but that association was anachronistic. Pythagoras, who had flourished seven centuries earlier, left behind no writings, and his followers had disappeared by the time of Aristotle. Whatever lore they had in arithmetic, geometry, and music had been assimilated into general intellectual culture and dissociated from these extinct Pythagoreans: the pseudo-Pythagorean texts composed between the time of Aristotle and the founding of the Roman Empire contain very little mathematical symbolism.[1]

With Nigidius Figulus (d. 45 BCE) began a movement to recover and lionize Pythagoras and his ideals. That movement, which we call neo-Pythagoreanism, was a literary or cultural ideal that claimed, among other things, that Pythagoras was the fountainhead of philosophy (mainly Plato's) and of number symbolism (its actual origins notwithstanding).[2] As Christianity developed, neo-Pythagoreanism became more widespread. In the first century CE the philosopher Moderatus of Gades wrote ten or eleven books of Pythagorean lore. Around the same time, Apollonius of Tyana adopted the lifestyle and teachings of Pythagoras and become an itinerant holy man. In the second century, Nicomachus of Gerasa, Numenius of Apamea, and Theon of Smyrna all wrote mathematical and philosophical texts laced with Pythagorean lore.

The influence is seen in Philo of Alexandria (fl. early first c. CE), a Jewish writer and political leader who was called a Pythagorean in antiquity, in part because he used allegorical number symbolism extensively.[3] He is one of the earliest writers known to have systematically compiled a handbook of ancient number symbolism. This treatise is lost, but an idea of its contents can be

[1] Thesleff 1961, 1965.
[2] Kahn 2001:88–93.
[3] Runia 1995.

gathered from many passages in Philo's extant writings.[4] His arithmology is significant for this study because it shows that number symbolism pervaded Jewish culture, which Christianity then shared, as much as Greco-Roman.

A good example is found in one of Philo's best-known texts, *On the Creation*, in which he argues for the intellectual coherence, even superiority, of Moses' account of the creation. Near the beginning of the treatise Philo explains why God is said to have created the world in six days. It is not as if God, who conceived and executed everything all at once, needed the extra time. Rather, in those six days God supplied order and rank to created beings. According to the laws of the nature of numbers, six is the number most conducive to begetting. Philo explains:

> For it is the first perfect number after the monad, equal to its parts and composed by them (half is a triad, a third is a dyad, a sixth is a monad). And it is, so to speak, male and female, fitted together by the power of each. For in things that exist odd is male and even is female. So the beginning of odd numbers is the triad, and of even numbers the dyad, but the power of both is the hexad. For the world, being the most perfect of created things, was put together in accordance with the perfect number, the hexad. And since it was about to have in itself things created from copulation, it had to be fashioned in accordance with a mixed number, the first even-odd. It was to encompass the form of the male, who sows the seed, and that of the female, who receives offspring.[5]

In this extract Philo presents several number symbols, some of which he assumes his readers know and accept. The first pertains to six. In antiquity, perfect numbers were those equal to the sum of their factors (including the number one). Hence six, whose factors are one, two, and three, is a perfect number. The second symbol pertains to the numbers two and three. All numbers were considered to have gender: even numbers were female, and odd numbers,

[4] Philo's lost treatise *On Numbers* is reconstructed by Stähle 1931; a fragment preserved in Armenian is found in Terian 1984. The tradition of handbooks of number symbolism began possibly with Posidonius (ca. 135–ca. 51 BCE). See Robbins 1921.
[5] Philo *On the Creation of the World* 13–14: τῶν τε γὰρ ἀπὸ μονάδος πρῶτος τέλειός ἐστιν ἰσούμενος τοῖς ἑαυτοῦ μέρεσι καὶ συμπληρούμενος ἐξ αὐτῶν, ἡμίσους μὲν τριάδος, τρίτου δὲ δυάδος, ἕκτου δὲ μονάδος, καὶ ὡς ἔπος εἰπεῖν ἄρρην τε καὶ θῆλυς εἶναι πέφυκε κἀκ τῆς ἑκατέρου δυνάμεως ἥρμοσται· ἄρρεν μὲν γὰρ ἐν τοῖς οὖσι τὸ περιττόν, τὸ δ' ἄρτιον θῆλυ· περιττῶν μὲν οὖν ἀριθμῶν ἀρχὴ τριάς, δυὰς δ' ἀρτίων, ἡ δ' ἀμφοῖν δύναμις ἑξάς. ἔδει γὰρ τὸν κόσμον τελειότατον μὲν ὄντα τῶν γεγονότων κατ' ἀριθμὸν τέλειον παγῆναι τὸν ἕξ, ἐν ἑαυτῷ δ' ἔχειν μέλλοντα τὰς ἐκ συνδυασμοῦ γενέσεις πρὸς μικτὸν ἀριθμὸν τὸν πρῶτον ἀρτιοπέριττον τυπωθῆναι, περιέξοντα καὶ τὴν τοῦ σπείροντος ἄρρενος καὶ τὴν τοῦ ὑποδεχομένου τὰς γονὰς θήλεος ἰδέαν. In this book all translations not otherwise credited are my own.

starting with three, were male. (The number one was frequently considered androgynous.) The gendering of number was both extensive and old, going back to Aristotle and the Pythagoreans.[6] These assumptions explain Philo's observation that six is the first number born from the multiplicative union of male and female numbers. So it makes sense that the creation happened in six days: no other number better typifies the fertility God implemented in the natural order. Philo builds upon both of these number symbols, one illustrating perfection and the other sexual generation, to argue that to be perfect and productive, creation had to have occurred in six days. Philo uses number, both here and throughout his writings, to unite seemingly disparate worlds.[7]

Philo's number symbolism was not exceptional. Flourishing shortly after him, Plutarch of Chaeronea used it just about as frequently, and in a similar fashion. In the sixties CE, Plutarch studied with Ammonius, an Alexandrian Platonist for whom mathematics and number symbolism were central. Ammonius' arithmological influence is seen throughout Plutarch's diverse oeuvre, especially in two treatises, now lost, devoted exclusively to numbers: *Whether the Odd or Even Number Is Better* and *Monads*.[8] But he shows contradictory attitudes toward number symbolism: in some passages he revels in it, and in other places he ridicules it.

This ambivalence is seen in *The E at Delphi*, which is styled as a Socratic or Aristotelian dialogue. Plutarch includes himself and Ammonius, his former teacher, as two of the six named participants. The subject at hand is the meaning behind an "E" inscribed at Delphi. The various participants note that the inscription could be read as a number or a letter, and either reading could be interpreted in different ways. One by one the participants offer seven possible explanations for why the letter was inscribed on the temple.

The first answer, proposed by Plutarch's brother Lamprias, takes the E as a numeral, which he says represents the five wise men of Greece: Khilon, Thales, Solon, Bias, and Pittakos, all of whom are reported to have met at Delphi, where they agreed to consecrate the letter in honor of their number (Plutarch of Chaeronea *The E at Delphi* 3 [385d–386a]). Ammonius dismisses the suggestion, as does the participant who offers the next explanation, that the E, being the second of seven vowels, represents the second of the seven planets—the sun, which followed the moon in ancient cosmology—and therefore Apollo (4 [386a–b]). This suggestion too is dismissed.

The third and the fourth hypotheses explain the E (in Plutarch's time it was called *ee*, not *epsilon*), according to the way it was spoken, εἰ. So one participant

---

[6] Sources and analysis in Burkert 1972:33–34.
[7] For other examples, see refs. in nn. 1–2 above.
[8] Lamprias's catalogue, nos. 74, 163.

says that the E represents either the word 'if' (εἰ), the keyword used to discover from the god the outcome of a future endeavor; or the 'if' governing the optative mood, to indicate wishes or prayers (4–5 [386b–d]). Another claims that the 'if' indicates the force of syllogistic logic (6 [386d–387d]).

The next two participants return to the possibility that E is a numeral. An Athenian suggests that it symbolizes the number five, "the pemptad" (7 [387d–f]).[9] Plutarch eagerly takes over the idea and pursues a lengthy excursus on the mathematical and symbolic excellence of the number five. He notes that five is the sum of the first odd and even numbers, and is therefore called 'marriage' (8 [387f–388c]). Five is called 'nature,' since when multiplied by itself the product's final digit is always five; when five is multiplied by any other number it results in either the decad or itself; and this behavior resembles nature, which returns either to itself or to perfection (8 [388c–e]). But what does this have to do with Apollo, the god at Delphi, who was traditionally associated with the numbers one and seven, not five? Plutarch offers a convoluted explanation that involves Dionysus, the harmony found in poetical measures, and finally the ratio of three to nine, seen in the relation between the creation and the conflagration (9 [388e–389c]).

As if entangled in his own obscurity, Plutarch leaves this line of thought unresolved, and reverts to his original explanation, that the number five is native to divinity because it generates either itself, like fire, or perfection, like the universe. So, he goes on, five also appears frequently in music, in the interval of the fifth (literally διὰ πέντε), one of the five basic intervals (10 [389c–f]). Furthermore, Plato affirmed there to be five worlds, Aristotle taught five elements, and there are five fundamental geometrical shapes in the *Timaeus*.[10] Plutarch associates the five senses with the five elements, and he refers to Homer, who divided the cosmos into five parts (12–13 [390b–c]). He appeals to the sequence point, line, plane, and solid, and argues for its continuation to a fifth level, the soul, which in turn naturally separates into five parts. There are five classes of living things in the world: gods, daemons, heroes, people, and beasts (13 [390c–f]).[11] Five is the sum of the first two squares, provided that one is willing to take the monad as a square (14 [390f–391a]). Plato posits five chief principles, causes, and categories. Therefore, Plutarch argues, the Delphic inscription was set up to anticipate Plato's doctrine (15 [391a–d]). The crowning point in this rambling

---

[9] The thought is repeated in Plutarch *The Obsolescence of Oracles* 36 (429d). See also Plutarch *Isis and Osiris* 56 (374a).

[10] Plutarch *The E at Delphi* 11 (389f–390a), a meditation expanded in his *Obsolescence of Oracles* 32–33 (427a–428b).

[11] The soul was normally trisected in the Platonic tradition, and, indeed, in Plutarch's other writings. Dillon 1996:194.

encomium—now longer than all the previous five explanations combined—is a riddle about why, when the priestess is led to the Prytaneum, five sortitions are first performed. One of the participants, Nikandros, warns that the reason should not be uttered. "Until such time as we become holy men," Plutarch answers, smiling, "and God grants us to know the truth, this also shall be added to what may be said on behalf of the Five."[12]

Ammonius, offering the seventh and final explanation, responds skeptically to Plutarch's praise. He notes that just about any number's praises could be sung, especially Apollo's native number seven (17 [391e–392a]). For him, the more plausible explanation for the E is "you are" (εἶ). For the rest of the treatise, this seventh and final explanation deals with what it means to ascribe eternal, unchanging being to Apollo, given our own fleeting, fluctuating condition (18–21 [392a–394c]).

The length of the sixth explanation shows that Plutarch was fascinated with numbers and their symbolism. But the philosophical heft of the seventh makes it difficult to determine how seriously he regarded them. The mathematical interpretation of the E is the longest, but it is trumped by Ammonius' ontological explanation. Of all the participants, Ammonius, the mathematician and Pythagorean, would be the one expected to emphasize the number symbolism of the E. At the end of *The E at Delphi*, the reader is left with no definite indication as to which explanation is most to be trusted.

In his study of Plutarch, Robert Lamberton notes how difficult it is to interpret the Delphic dialogues: "deliberate but coy self-portraits" where "Plutarch remains a very elusive presence."[13] In his reading, the dialogues dramatize inquiry and keep a single, dominating explanation at arm's length, emphasizing instead the importance of dialogue and the pursuit of truth. This idealization of the pursuit—not the attainment—of truth is symbolized by the setting in Delphi, where the oracle speaks in riddles easily misinterpreted, and where the priests may not divulge the mysteries. Plutarch, himself a Delphic priest when he wrote the dialogue, would never have revealed the secrets of the priesthood, so to search the discourse for the correct answer is a fool's errand.[14] But if the arithmological explanation of the E is merely one installment in the search for truth, then why has Plutarch dwelt on it and at such length? What does he intend the reader to do with all this numeric lore?

One obvious answer, in light of Lamberton's thesis, is that Plutarch wants his readers to be entertained, to have their repertoire of knowledge expanded.[15]

---

[12] *The E at Delphi* 16 (391d–e): "οὐκοῦν" ἔφην ἐγὼ μειδιάσας "ἄχρι οὗ τἀληθὲς ἡμῖν ὁ θεὸς ἱεροῖς γενομένοις γνῶναι παράσχῃ, προσκείσεται καὶ τοῦτο τοῖς ὑπὲρ τῆς μεμπάδος λεγομένοις."

[13] Lamberton 2001:5.

[14] Lamberton 2001:26, 149, 156–158.

[15] Dillon 1973:190.

Such a casual attitude toward number symbolism is illustrated best in *Table Talk*, a lengthy collection of festive after-dinner conversations. There, number symbolism crops up in many of these discussions with an air of sport and riddle, and the points are treated as intellectual curiosities, nothing more. But in other treatises Plutarch encourages his readers to use number symbolism to understand the world. This is best seen in *Isis and Osiris*, an analysis of traditional Egyptian religion and mythology.

Some of the number symbolism behind Egyptian mythology Plutarch finds ridiculous but fascinating. For example, Typhon, when hunting in the light of the moon, found and chopped up Osiris' corpse into fourteen parts. That this dismemberment relates to the lunar phases Plutarch has no doubt.[16] But he goes on to report the notion that cats give birth to successively larger litters—one, two, three, all the way up to seven kittens, thereby giving birth to twenty-eight in all.[17] He admits that the story is far-fetched, but he finds it uncanny that cats' eyes dilate when the moon is full.[18]

Most of the symbolic numbers in *Isis and Osiris* Plutarch treats as key interpretive tools for religious myth. For example, he claims that the Egyptians assigned numbers to gods and thereby inspired the Pythagoreans. Apollo is the monad, Artemis is the dyad, Athena is the hebdomad, and Poseidon is the first cube (eight)—familiar Greek motifs that were originally part of Egyptian religion.[19] Plato's nuptial triangle, consisting of sides of lengths three, four, and five, also derives from the Egyptian myth of Isis and Osiris.[20] The side of length three symbolizes the male, the base of four the female, and the hypotenuse their offspring. These three sides correspond to Osiris as the source (ἀρχή), Isis as the receptacle (ὑποδοχή), and Horus as the completion (ἀποτέλεσμα). The number symbolism follows a pattern of interpretation comparable to Philo's explanation of the six days of creation: three (Osiris) is the first odd number and is perfect; four (Isis) is a square with sides of the (first) even number two; and five (Horus) is likened to both its father (three) and mother (two). In this allegory, five symbolizes not only marital union but the perfect offspring of that union.[21]

---

[16]   *Isis and Osiris* 18 (358a), 42 (368a).

[17]   *Isis and Osiris* 63 (376e).

[18]   Compare Galen, who more forcefully argues against the Pythagoreans, who theorized that odd-numbered days, because they are male, induce a major change (κρῖσις) in a sick patient (*De diebus decretoriis* 922–924). Galen attributes this power to the moon and its rhythms.

[19]   Plutarch *Isis and Osiris* 10 (354–355).

[20]   Plato *Republic* 546b; Plutarch *Isis and Osiris* 56 (373f–374a).

[21]   Cf. Philo *On the Contemplative Life* 65–66, which claims that the number five is "the most natural" (φυσικότατος) number because it is drawn from the orthogonal triangle that is the "source of generation of the universe" (ἀρχὴ τῆς τῶν ὅλων γενέσεως). For another extended application of the orthogonal triangle to generation and copulation (but without direct reference to five as 'marriage'), see the scholia on Homer, set D, 19:119. There, the triangle is used to explain why

Although he treats numbers anthropologically, as devices for understanding human myths of the world, Plutarch also handles them with some reverence, as portals into the realm of the gods. This harmonizes with his views on symbolism in general. When approaching the stories of the gods, Plutarch avoids the two extremes of superstition and atheism.[22] His middle way is to adopt a philosophical and pious attitude to the various customs of the world, and thereby find the truth embedded in symbols. Plutarch says that just as the planets and elements, though named differently, are held in common, so too people may give the gods different names, but they nevertheless share the same ones.[23] Behind all these naming systems, behind all the various religious symbols, is a single reason (λόγος) and providence that orders and guides everything.

Number symbolism was part of that single reason and providence for both Philo and Plutarch. In the first century, to engage piously with symbols was to grapple with the truth. Number symbols could explicate religious mythology and thereby persuade readers to embrace a particular religious world view. They could provide a gossamer portal to the harmonies of the world or to the divinity that pervades it. But number symbols could also be abused or misinterpreted. They could be tools of superstition, such as in Plutarch's numbers concerning cats. Or they could be contested, just as the six days of creation could be a sign of either the inferiority or the superiority of the God of the Bible. Numbers were important for interpreting the world, and for persuading others to embrace that interpretation.

Monism, Dualism, and Pluralism

It is well known that the earliest Greek philosophers sought a single, original principle in the world, one ἀρχή to account for the existence of all things. In the sixth century BCE the Ionians grounded this principle in the material realm. Thales identified the ἀρχή as water; Anaximander as the indefinite (τὸ ἄπειρον); Anaximenes as air. This impulse to the monistic shares an affinity with even older Orphic cosmogonies, which identify night, Khronos, or water as the sole progenitor of the universe.[24]

Alongside the Ionian monistic impulse was a cultural tendency toward pluralism, widespread in myth. In Hesiod, the two beings Earth and Sky are a single, commingled original entity that undergoes a split, to engender Chaos.

---

infants are viable in the seventh and ninth months, but not in the eighth, a common belief in the ancient world.

[22] *On Superstition* and *Isis and Osiris* 11 (355c–d), 67 (378a). Cf. Gwyn Griffiths 1970:100–101, 291.

[23] Plutarch *Isis and Osiris* 67 (377e–f).

[24] Kirk et al. 1983:22–33.

One strain of Orphic mythology has Water paired with Matter/Earth as the beginning of all things. Pherecydes held that there were three original eternal beings: Zas, Khronos, and Khthonie, conceived of both mythologically and in cosmological abstraction.[25] All three of these sources date to the sixth century BCE, and all three attest to a shared sense of the origins of the universe—an original plurality full of strife and struggle, a cosmic drama that explains the conflicts found in succeeding generations, those of gods, titans, heroes, and humans.

This pluralism is also reflected in other strains of early Greek philosophy, perhaps in reaction to Ionian monism. Heraclitus, for example, saw the cosmos as a theater for the strife of opposites, a strife held in tense unity by the logos, symbolized by fire. Empedocles posited as first principles the four elements: fire, air, water, and earth—four roots of the universe whose intermingling was regulated by love and strife. Empedocles' contemporary Anaxagoras taught that an infinite number of material elements always existed, in mixture, in a kind of dualism of mind and matter. And the atomists, most notably Democritus, held to a dualism *cum* pluralism: the two basic principles of the universe—the full and the void—plus the infinite number of atomic elements.[26] The earliest Pythagoreans were also pluralists.[27]

Under the influence of Parmenides, monism nearly vanished from Greek philosophy. This is not because he was a dualist, since it is unclear how many principles he taught.[28] But Parmenides' central argument—that existence excludes becoming and change—challenged subsequent philosophers to provide an account of the world that satisfied both his criteria for existence and our experience of a world of change. If the one original principle truly existed, it could not change. Some other principle would be required to explain the change we observe.

In light of Parmenides' challenge, it is not surprising that Plato espoused a kind of dualism, but he cloaks it in a language of shifting terminology and metaphor. For instance, in the *Timaeus* the cosmogony begins with two entities: the demiurge and the receptacle. The primal existence of both is assumed, and not explained. In *Philebus*, Plato (through Socrates) wrestles with the relationship between the one and the many, and again he presumes but never explains the

---

25 Kirk et al. 1983:34–41, 24–26, 56–57. See also Burkert 1972:38–39.
26 Kirk et al. 1983:414–415.
27 For more on Pythagorean pluralism and its relationship to Parmenides see Burkert 1972:32–35, 48–49, and Huffman 1993:23.
28 Patricia Curd has challenged the long-standing consensus that Parmenides was a material monist (2005:xvii–xxix). The only known pre-Platonic monist after Parmenides was Diogenes of Apollonia (Curd 2005:131). For a more complete account of the turn from monism to pluralism via Parmenides, see Rist 1965.

preexistence of both principles. He offers instead a process by which two opposites come to coexist in unity. Hence the third "type" in *Philebus*, the combined synthesis of the definite and indefinite, the classical model of balance between opposites. The notion of two original principles appears in the *Republic*, where the realm of ideas is subordinated to a single principle, the good. In other dialogues the good stands in opposition to a second principle, and this second principle is itself conceived of as a pair of contrasts, for instance "the great and the small."[29] This last leitmotif is developed by Aristotle, who says that Plato embraced the one and the indefinite dyad (i.e. the "great and small") as his principles. The one is the essence of the forms, and the dyad (also called 'great-and-small' and likened to other contrasts) is their matter.[30]

Some of the names for the principles—one, many, indefinite dyad—show that Plato frequently thought of the highest, immaterial world in numerical terms. This is no surprise, because he treated the entire universe, not just its immaterial component, as a chain of numbers. According to Aristotle (*Metaphysics* a6), Plato's ideas exist in an upper, incorporeal realm of forms above the sensible world, and corporeal objects exist by participating in these higher ideas. The ideas and the material world are linked by an intermediate realm of numbers, distinct from the lower, material realm because they are unchanging and eternal, and from the upper, ideal realm because they are multiply instantiated. These numbers, too, come from above, through the participation of the great-and-small with the one.[31] Thus numbers are a layer between the noumenal realm and the material, providing Plato's system with three levels, not the traditional two. The position departs from classical Pythagorean number theory: the Pythagoreans claimed that numbers were corporeal, and that sense-perceptible things are, or are made of, numbers.[32] But for Plato, number is rooted in the upper realms, not the lower material world.

Around the first century BCE a new trend emerged in Platonist metaphysics. On the one hand classical dualism was challenged, as a group of philosophers began to champion a pure monism.[33] But where one of Plato's positions was discarded, another was magnified. The monistic principle these new

---

[29] Burkert 1972:17, and Ross 1924:169–171.

[30] Outlined at *Metaphysics* 987b14–29, with further refs. and discussion in Burkert 1972:21–22.

[31] Ross 1924:166–168 for references and discussion. See also Annas 1976:13–21, on the possibility that the intermediate mathematicals were introduced after Plato.

[32] Burkert 1972:31–34. Although Aristotle misrepresents the Pythagorean tradition in key places, he seems to be accurate here. For according to testimony independent of Aristotle, the Pythagorean Philolaus championed not numbers but limiters and unlimiteds as his highest principles. And even these principles are corporeal, not immaterial; see Huffman 1993:37–53.

[33] On the shift from older, dualist Pythagoreanism to the innovative monist strain see Dillon 1996:344, 373, 379; Armstrong 1992:34–41; Kahn 2001:97–99; and Thomassen 2000:3–4.

philosophers envisioned, nothing like the material ἀρχαί of the Ionian philosophers, sat atop an edifice of three or more levels of immaterial reality. And the levels related to each other primarily in arithmetical terms. So Plato's impulse to think mathematically about the world and to stratify the world into multiple levels of reality was adopted and expanded.

This trend is central to the early Christian theology of arithmetic, so here we must slow down. Although the later effects are quite evident, the original sources are scarce, and tersely reported. We have far fewer texts of Hellenistic philosophy (treatises written after Aristotle but before Plotinus) than we do for other periods, and it is easy to allow the voluminous writings of Plutarch and Philo—who did not participate in this trend—to drown out other writers from the period who were equally prolific but not lucky enough to have their works preserved. To linger on these fragmentary sources is important, because entire metaphysical systems are condensed and summarized in only a few sentences, often in obscure Greek where every word counts.

The earliest fragment attesting to this trend is found in Alexander Polyhistor's *Successions of the Philosophers* (ca. first c. BCE), which quotes from a text called *Pythagorean Memoirs* (itself of unknown date):

> "The principle (ἀρχή) of all things is a monad. And from the monad the indefinite dyad exists, like matter for the monad, its cause. From the monad and the indefinite dyad come the numbers, and from the numbers the points. From them come lines, out of which come planar figures. From planes come solid figures, and from these, sense-perceptible bodies, from which come the four elements—fire, water, earth, air."[34]

The text goes on to describe the composition of the Earth in terms reminiscent of the cosmogony in Plato's *Timaeus*.

We do not have enough of Alexander's original text—indeed, we know hardly anything about Alexander himself—to know what he made of this system. But the quotation above shows that in the first century BCE there was circulating a Pythagorean cosmology with three innovative features. First, it emphasized the monadic origin of the world: only the monad is preexistent; the dyad exists because of the monad. Second, it multiplied the number of immaterial levels— seven realms unfold, one from the other, before corporeality is achieved. And

---

[34] ἀρχὴν μὲν τῶν ἁπάντων μονάδα· ἐκ δὲ τῆς μονάδος ἀόριστον δυάδα ὡς ἂν ὕλην τῇ μονάδι αἰτίῳ ὄντι ὑποστῆναι· ἐκ δὲ τῆς μονάδος καὶ τῆς ἀορίστου δυάδος τοὺς ἀριθμούς· ἐκ δὲ τῶν ἀριθμῶν τὰ σημεῖα· ἐκ δὲ τούτων τὰς γραμμάς, ἐξ ὧν τὰ ἐπίπεδα σχήματα· ἐκ δὲ τῶν ἐπιπέδων τὰ στερεὰ σχήματα· ἐκ δὲ τούτων τὰ αἰσθητὰ σώματα, ὧν καὶ τὰ στοιχεῖα εἶναι τέτταρα, πῦρ, ὕδωρ, γῆν, ἀέρα· (quoted in Diogenes Laertius [fl. third c. CE] *Lives of the Philosophers* 8.25).

third, these seven regions are primarily mathematical—arithmetic precedes geometry, and within arithmetic monad precedes dyad, both of which precede numbers. The world is described numerically, with the monad advancing step by step to fill up the material universe. Thus, although the author of the *Pythagorean Memoirs* draws from motifs found in the cosmogony of the *Timaeus*, he develops a position that differs from classic Platonism and early Pythagoreanism—indeed from any form of ancient Greek philosophy.

Eudorus (fl. late first c. BCE), who lived shortly after Alexander, also discusses the Pythagoreans. Here are his words, preserved by the sixth-century philosopher Simplicius:

> "It must be stated that according to the higher account the Pythagoreans call the one the principle (ἀρχή) of all things, but according to the other account there are two principles of things that are perfected—the one and the nature opposite to it. Out of everything that can be considered according to opposites that which is noble is subordinated to the one; and that which is base, to the nature set in opposition to it. Thus, these are in no way principles, by these men's account. For if there is one principle for these things, [but] another for those, they are not common principles for all things, as is the case with the one.... Wherefore ... they say that the one is the principle of all things in a different manner, as if matter and all existent things come into existence from it. This is the upper god."[35]

After a brief interlude the quote continues:

> "I [Eudorus] say that those in Pythagoras' circle abandoned the one as the principle of all things, and in a different manner brought in the two highest elements. They call these two elements by various names. For one of them is called ordered, defined, knowable, male, odd, right, light; its opposite, disordered, indefinite, unknowable, female, even, left, darkness. Thus on one hand they call the one a principle, but on the other hand [they call] the one and the indefinite dyad elements—[the]

---

[35] "κατὰ τὸν ἀνωτάτω λόγον φατέον τοὺς Πυθαγορικοὺς τὸ ἓν ἀρχὴν τῶν πάντων λέγειν, κατὰ δὲ τὸν δεύτερον λόγον δύο ἀρχὰς τῶν ἀποτελουμένων εἶναι, τό τε ἓν καὶ τὴν ἐναντίαν τούτῳ φύσιν. ὑποτάσσεσθαι δὲ πάντων τῶν κατὰ ἐναντίωσιν ἐπινοουμένων τὸ μὲν ἀστεῖον τῷ ἑνί, τὸ δὲ φαῦλον τῇ πρὸς τοῦτο ἐναντιουμένῃ φύσει. διὸ μηδὲ εἶναι τὸ σύνολον ταύτας ἀρχὰς κατὰ τοὺς ἄνδρας. εἰ γὰρ ἡ μὲν τῶνδε ἡ δὲ τῶνδέ ἐστιν ἀρχή, οὐκ εἰσὶ κοιναὶ πάντων ἀρχαὶ ὥσπερ τὸ ἕν." καὶ πάλιν "διό, φησί, καὶ κατ᾽ ἄλλον τρόπον ἀρχὴν ἔφασαν εἶναι τῶν πάντων τὸ ἕν, ὡς ἂν καὶ τῆς ὕλης καὶ τῶν ὄντων πάντων ἐξ αὐτοῦ γεγενημένων. τοῦτο δὲ εἶναι καὶ τὸν ὑπεράνω θεόν" (Simplicius *Commentary on Aristotle's "Physics"* 9.181.10–19).

one exists recurrently, so both are principles. And clearly, one of the ones is the principle of all things, but the other one is the dyad's opposite, what they call a monad."[36]

Most scholars dealing with this passage believe that Eudorus is describing a single Pythagorean account of the principles.[37] In my opinion, Eudorus is critically comparing two competing systems. According to what he calls the loftier version, the Pythagoreans held the one to be the principle of all things. According to their second version there are two principles, which preside over things considered opposites. Eudorus criticizes this second account for its incompleteness. The lower pair of principles cannot properly be deemed principles, as the Pythagoreans claim, since they generate not everything but only opposites. In contrast, according to their monistic version, not just some but all matter and all things exist thanks to the one, the so-called higher god. Eudorus accuses the Pythagoreans of abandoning a monistic principle and introducing a dyadic one, and of adopting new terminology, preferring the term 'elements' (στοιχεῖα) to 'principles' (ἀρχαί). He argues that because the term 'one' is used in both schemes, they should both describe the same entity—the one. So if the one is a principle in the first system, it should be the same in the other. But this runs contrary to their terminology and results in a serious inconsistency: the one is given contradictory functions.

This testimony shows that both monist and dualist versions of Pythagoreanism were then circulating, and that Eudorus preferred the monist one as being truer and closer to Pythagoras' original teaching. This is a testament to how thoroughly this new strain of Platonism was rewriting the legacy of Pythagoras and Plato. Eudorus' prejudice is confirmed by another of his preserved fragments, in which he emends the text of Aristotle's *Metaphysics* to have Plato say that the one is responsible for the existence of the forms and matter, thus subordinating matter to the one.[38] So Eudorus considers monism to be the superior philosophy, echoing the monism found in the *Pythagorean Memoirs*.

---

[36] "φημὶ τοίνυν τοὺς περὶ τὸν Πυθαγόραν τὸ μὲν ἓν πάντων ἀρχὴν ἀπολιπεῖν, κατ' ἄλλον δὲ τρόπον δύο τὰ ἀνωτάτω στοιχεῖα παρεισάγειν. καλεῖν δὲ τὰ δύο ταῦτα στοιχεῖα πολλαῖς προσηγορίαις· τὸ μὲν γὰρ αὐτῶν ὀνομάζεσθαι τεταγμένον ὡρισμένον γνωστὸν ἄρρεν περιττὸν δεξιὸν φῶς, τὸ δὲ ἐναντίον τούτῳ ἄτακτον ἀόριστον ἄγνωστον θῆλυ ἀριστερὸν ἄρτιον σκότος, ὥστε ὡς μὲν ἀρχὴ τὸ ἕν, ὡς δὲ στοιχεῖα τὸ ἓν καὶ ἡ ἀόριστος δυάς, ἀρχαὶ ἄμφω ἓν ὄντα πάλιν. καὶ δῆλον ὅτι ἄλλο μέν ἐστιν ἓν ἡ ἀρχὴ τῶν πάντων, ἄλλο δὲ ἓν τὸ τῇ δυάδι ἀντικείμενον, ὃ καὶ μονάδα καλοῦσιν" (Simplicius *Commentary on Aristotle's "Physics"* 9.181.22–30).

[37] My reading of Eudorus differs from that of Rist (1962:391–393), who sees Eudorus as having muddled up a single Pythagorean system, the same system Alexander Polyhistor reports. See also Dillon 1996:115–135; Kahn 2001:97–98; Trapp 2007:351–355; Bonazzi 2007; and Staab 2009.

[38] Rist 1962:394.

Arithmetic also plays a key role in the systems Eudorus reports. The terms 'one,' 'monad,' and 'dyad' recur throughout, and Eudorus highlights how the one functions numerically in the metaphysical scheme. His critique is built upon the notion that, on its own, the one is capable of generating all beings; but when treated as a peer of the indefinite dyad its function is incomplete, responsible for only half of a limited set of objects. Both Eudorus and his loftier Pythagorean system prefer an account of the universe where the one is responsible for the generation of all things.

The Pythagorean themes advanced by Eudorus and the *Pythagorean Memoirs* are confirmed by Moderatus, who flourished in the first century CE and who, like the other two, collected and commented on Pythagorean lore.[39] In two key passages, Moderatus outlines two metaphysical systems, the first belonging to the Pythagoreans, the second to Platonists. Moderatus' text, like Eudorus', is preserved by Simplicius, but this time via Porphyry's treatise *On Matter*. Moderatus' account falls in two sections. This first is somewhat paraphrastic, with explanatory comments from either Simplicius or, more likely, Porphyry. Simplicius states:

> And the first among the Greeks to have apparently held this theory about matter [that the differences from materiality to immateriality are marked by invisible, immeasurable criteria] are the Pythagoreans, and after them Plato, as Moderatus also relates. For he declares that according to the Pythagoreans, the first one is shown to be beyond existence and every essence. But the second one (which is the truly existent and object of intellection) he says is the forms. And the third (which is that of the soul) participates in the one and the forms, and the final nature [that derives] from it, which is of sense perceptibles, does not participate, but is arranged in accordance with their reflection. The matter that is in [the sense perceptibles], of nonbeing first existing in quantity, is [termed] 'eclipse-shadow' and [is] yet even further inferior and [derives] from it.[40]

---

[39] For a fuller account of Moderatus, see Dillon 1996:344–351 and Kahn 2001:105–110.

[40] (230.34) Ταύτην δὲ περὶ τῆς ὕλης τὴν ὑπόνοιαν ἐοίκασιν ἐσχηκέναι πρῶτοι (35) μὲν τῶν Ἑλλήνων οἱ Πυθαγόρειοι, μετὰ δ' ἐκείνους ὁ Πλάτων, ὡς καὶ Μοδέρατος ἱστορεῖ. οὗτος γὰρ κατὰ τοὺς Πυθαγορείους τὸ μὲν πρῶτον ἓν ὑπὲρ τὸ εἶναι καὶ πᾶσαν οὐσίαν ἀποφαίνεται, τὸ δὲ δεύτερον ἕν, ὅπερ ἐστὶ (231.1) τὸ ὄντως ὂν καὶ νοητόν, τὰ εἴδη φησὶν εἶναι, τὸ δὲ τρίτον, ὅπερ ἐστὶ τὸ ψυχικόν, μετέχειν τοῦ ἑνὸς καὶ τῶν εἰδῶν, τὴν δὲ ἀπὸ τούτου τελευταίαν φύσιν τὴν τῶν αἰσθητῶν οὖσαν μηδὲ μετέχειν, ἀλλὰ κατ' ἔμφασιν ἐκείνων κεκοσμῆσθαι, τῆς ἐν αὐτοῖς ὕλης τοῦ μὴ ὄντος πρώτως ἐν τῷ ποσῷ ὄντος οὔσης σκίασμα καὶ ἔτι μᾶλλον ὑποβεβηκυίας καὶ ἀπὸ τούτου (Simplicius *Commentary on Aristotle's "Physics"* 9.230.34–231.5). In my translation I have supplied in square brackets the text to be understood from the Greek text or the context. Parts of the translation enclosed by parentheses correspond to Greek text that I regard as the

This passage is well known among scholars of intellectual history. Ever since E. R. Dodds wrote his seminal 1928 article, Moderatus has been widely celebrated for providing the earliest datable version of what would become the standard metaphysical edifice of Neoplatonism, inspired by the reading of Plato's *Parmenides*: a series of ones in descending tiers—from transcendence to intellection to soul and finally to sense perception.[41] Three levels of immaterial reality transcend the material plane, completely unattached except by imitation. The material world merely reflects the upper realms, unlike the soulish realm, which participates in the higher levels. The sense-perceptible world is but an image, a shadow of the incorporeal matter that resides above it.

This highly compressed fragment is worth reading again. Although it is a sludge of jargon, it tells a complex metaphysical story. The first half outlines the relationships of a hierarchy of three ones. The second half explains the material world, in terms both of nature and of matter, which derive from the third one. The physical world is both a decorative impression of the upper ones, and the first existential instantiation of nonexistence, the result of unity's entering the dimension of quantity. The phrase 'eclipse-shadow' (σκίασμα) implies a technical term familiar to Moderatus' audience. The term alludes to Plato's allegory of the shadows cast by a fire on a cave wall, but more directly likens the material world to the shadow cast by the earth on the moon, implying that nonexistence (μὴ ὄν), the act of first existing (πρότως ὄν), and quantity (ποσότης) were likened, in the original metaphor, to the sun, sunlight, and earth. Like the two Pythagorean systems described earlier, Moderatus' Pythagorean system is dominated by monism, with the superexistential one presiding over lower monistic levels and bestowing on them its unitary constitution. The metaphysical edifice is built from the blocks of arithmetic, with the four levels of the universe emerging from unity through quantity into material reality.

At the beginning of the quote above, Simplicius promises Moderatus' account of what the Pythagoreans and Plato thought about matter. The first passage presents the Pythagorean account. Simplicius' second passage from Moderatus follows on the heels of the first, reporting the Platonist account. This time it is a direct quotation, again via Porphyry. Note the arithmetical terms 'unitary' and 'quantity':

And these things Porphyry has written about in the second book of *On Matter*, quoting the [comments] of Moderatus: " 'When the unitary logos

---

explanatory comments of Simplicius or Porphyry. For earlier literature see Dörrie and Baltes 1996:477, Tornau 2001, and Staab 2009:71–76.

[41] Dodds 1928.

wanted, as Plato says somewhere [*Timaeus* 29d7–30a6], to have the generation of beings constituted from itself, it made room for quantity by its own privation, by depriving [quantity] of all its *logoi* and forms. He called this a quantity shapeless, undifferentiated, and without outline, but receptive of shape, form, differentiation, quality, and everything of this sort.' He says, 'Plato was likely to have predicated of this quantity a great number of names, calling it all-receptive and formless and invisible and least capable of participating in the object of intellection and scarcely grasped by spurious reasoning and everything related to these [terms].

"'This very quantity,' he says, 'and that form that is thought of according to the privation of the unitary logos, which encompasses in itself all the *logoi* of beings, are paradigms of the matter belonging to bodies,' the very [matter] that he said both the Pythagoreans and Plato called quantity; not quantity as a form, but according to privation and loosening and extension and dispersion and because of its deviation from being. Because of these things matter appears to be evil, as if fleeing the good, and is apprehended by it and is not allowed to depart from its limits, as, on one hand, the extension receives and is limited by the logos of the form-magnitude, and as, on the other hand, the dispersion is made into a form by arithmetical distinction." Thus, matter is, by this account, nothing other than the deviation of sense-perceptible forms vis-à-vis the objects of intellection, [forms that] wander off from that realm and are dragged down toward nonexistence.[42]

---

[42] (231.5) καὶ ταῦτα δὲ ὁ Πορφύριος ἐν τῷ δευτέρῳ Περὶ ὕλης τὰ τοῦ Μοδεράτου παρατιθέμενος γέγραφεν ὅτι "βουληθεὶς ὁ ἑνιαῖος λόγος, ὥς πού φησιν ὁ Πλάτων, τὴν γένεσιν ἀφ' ἑαυτοῦ τῶν ὄντων συστήσασθαι, κατὰ στέρησιν αὐτοῦ ἐχώρησε τὴν ποσότητα πάντων αὐτὴν στερήσας τῶν αὐτοῦ λόγων καὶ εἰδ(10)ῶν. τοῦτο δὲ ποσότητα ἐκάλεσεν ἄμορφον καὶ ἀδιαίρετον καὶ ἀσχημάτιστον, ἐπιδεχομένην μέντοι μορφὴν σχῆμα διαίρεσιν ποιότητα πᾶν τὸ τοιοῦτον. ἐπὶ ταύτης ἔοικε, φησί, τῆς ποσότητος ὁ Πλάτων τὰ πλείω ὀνόματα κατηγορῆσαι "πανδεχῆ" καὶ ἀνείδεον λέγων καὶ "ἀόρατον" καὶ "ἀπορώτατα τοῦ νοητοῦ μετειληφέναι" αὐτὴν καὶ "λογισμῷ νόθῳ μόλις ληπτήν" (15) καὶ πᾶν τὸ τούτοις ἐμφερές. αὕτη δὲ ἡ ποσότης, φησί, καὶ τοῦτο τὸ εἶδος τὸ κατὰ στέρησιν τοῦ ἑνιαίου λόγου νοούμενον τοῦ πάντας τοὺς λόγους τῶν ὄντων ἐν ἑαυτῷ περιειληφότος παραδείγματά ἐστι τῆς τῶν σωμάτων ὕλης, ἣν καὶ αὐτὴν ποσὸν καὶ τοὺς Πυθαγορείους καὶ τὸν Πλάτωνα καλεῖν ἔλεγεν, οὐ τὸ ὡς εἶδος ποσόν, ἀλλὰ τὸ κατὰ στέρησιν καὶ παρά(20)λυσιν καὶ ἔκτασιν καὶ διασπασμὸν καὶ διὰ τὴν ἀπὸ τοῦ ὄντος παράλλαξιν, δι' ἃ καὶ κακὸν δοκεῖ ἡ ὕλη ὡς τὸ ἀγαθὸν ἀποφεύγουσα. καὶ καταλαμβάνεται ὑπ' αὐτοῦ εἰδητικοῦ μεγέθους λόγον ἐπιδεχομένης καὶ τούτῳ ὁριζομένης, τοῦ δὲ διασπασμοῦ τῇ ἀριθμητικῇ διακρίσει εἰδοποιουμένου. ἔστιν (25) οὖν ἡ ὕλη κατὰ τοῦτον τὸν λόγον οὐδὲν ἄλλο ἢ ἡ τῶν αἰσθητῶν εἰδῶν πρὸς τὰ νοητὰ παράλλαξις παρατραπέντων ἐκεῖθεν καὶ πρὸς τὸ μὴ ὂν ὑποφερομένων (Simplicius *Commentary on Aristotle's "Physics"* 9.231.5–27).

Many commentators on Moderatus presume that this second passage explains the first, when in fact they are independent. This is signaled in the preamble, and by Simplicius' quotation style (giving only the second quote a bibliographical reference). The difference is apparent too in the content. The second passage diverges from the first, not merely summarizing but expanding upon Plato's system and its interpretive tradition, to provide the rationale for an innovative account of the structure of the universe. Simplicius, or perhaps Porphyry, had extracted two quotes from Moderatus on the concept of quantity, and tried to relate them. But they resist such a close comparison.[43]

In this second passage, a distinct metaphysical system is described, from the top down. At the highest point is the unitary λόγος, an entity that possesses within itself all the forms and λόγοι of existent things—exactly how this upper-most entity is constituted and how it possesses an internal plurality of forms and λόγοι is not explained. But it desires things to come into existence, so by privation it creates a vacuum for quantity, the second-level entity.[44] Along with

---

[43] Moderatus had an affinity for Pythagorean lore, whether he agreed with it or not (Porphyry *Life of Pythagoras* 48). This second passage falls into three parts: a quote from Moderatus, an interjected explanation from Porphyry, and then a paraphrase of yet more of Moderatus' account. So although we can expect the first part to reflect Moderatus' words, the second and third parts are Porphyry's synthesis, drawn from who knows what other parts of Moderatus' text. Distinguishing Simplicius', Porphyry's, and Moderatus' voices is admittedly difficult. For those who know the text and have puzzled over it, here is my rationale: The clearest indicators for separating the three authors are verbs of speaking. The παρατιθέμενος of lines 5–6 (Diels 1882–1895:1.231) is one of Simplicius' favored terms for introducing extended quotations, which he tends to reproduce conscientiously. (See e.g. his quote of Geminus paraphrasing Posidonius at 291.21–292.31, which also begins with παρατίθησιν, and in which he inserts no φησι of his own.) Thus I regard line 7 onward as having no words written by Simplicius (ending, presum-ably, at the close of line 24, where Simplicius assesses the role of ὕλη). In that section, verbs of speaking appear at lines 7 (φησιν ὁ Πλάτων), 10 ([Πλάτων] ἐκάλεσεν), 12 ([Μοδέρατος] φησι), 13 (ὁ Πλάτων ... λέγων), 15 (φησι), and 18–19 (τοὺς Πυθαγορείους καὶ τὸν Πλάτωνα καλεῖν [ὁ Μοδέρατος] ἔλεγεν). The φησι at line 15 likely has Moderatus and not Plato as its subject, parallel to the φησι at line 12. (Thus the quotation marks in my translation, distinguishing Moderatus' voice from Porphyry's). The change from φησι (12, 15) to ἔλεγεν (18–19) is significant, for the latter not only changes the tense but introduces indirect discourse, a sign of Porphyry's sum-marizing or paraphrasing Moderatus (also in evidence in our first passage). Further, the clause in which ἔλεγεν appears (starting at line 18 with ἦν) governs the rest of the quote down through line 24, a single sentence (despite Diels's unnecessary terminal punctuation at line 21); and this long text comprises five explanatory clauses (18: ἦν, 19: οὐ ... ἀλλὰ, 20: δι' ἅ, 22: τῆς μὲν, 24: τοῦ δὲ). So lines 18–24 contain Porphyry's summary of Moderatus' system, drawn from texts not necessarily near the source of the quotation in lines 7–18. For other opinions on attribution, see Dodds 1928:138n3 and Tornau 2001:204–205n26.

[44] On the background of the concept of privation, see Tornau 2001:207–208; for its later use, Thomassen 2006:271–279. Cf. *Theology of Arithmetic* 9.5–6, where the dyad divides itself from the monad: πρώτη γὰρ ἡ δυὰς διεχώρισεν ἑαυτὴν ἐκ τῆς μονάδος. For scholarly opinions on the nature and function of the internal forms and λόγοι of the unitary λόγος, see Tornau 2001:209n37.

quantity is an entity called 'form,' whose existence is also unexplained.[45] These two entities become paradigms for corporeal matter, which is labeled by the Pythagoreans and Plato as "quantity." This quantity—not to be confused with the upper, ideal quantity mentioned here and at the end of Moderatus' Pythagorean report in the first passage—is a material entity that arises through a series of acts of estrangement. Its extension is guided and regulated by the proportionate interaction of ideal quantity (here called magnitude) and form. Its dispersion is structured by numbers. The four verbal nouns—privation, loosening, extension, and dispersion—describe four stages in the creation of corporeality, starting with void in the center and ending in the dispersed material world, a non-existence.[46]

The numerical language—'unitary logos,' 'quantity,' 'logoi' (implying mathematical ratios), and 'arithmetical distinction'—runs throughout Moderatus' Platonic system, presenting the origin of the world as if it were a cascade of numbers. The system may seem monist, but the constitution of the unitary logos is unclear. This entity has volition and, as Christian Tornau points out, rationality—which entails objects that are thought, and therefore plurality.[47] Within the unitary logos exist forms and λόγοι, which implies that dualism, pluralism, or even an attempt at pluralism-in-unity is at work. Thus, although Moderatus reports the monism of the Pythagoreans, it is unclear whether he was himself a monist.

Taking into account all three Middle Platonist authors—the *Pythagorean Memoirs*, Eudorus, and Moderatus—we see a new emphasis on metaphysics, an attempt to explicate what Plato left unsaid or unclear.[48] Multiple levels of reality are postulated, emanating from the top down or from the center outward in imitation of the generation of numbers. To describe the cosmogony, mathematical metaphors are essential. Against the traditional dualism of Hellenistic Platonists, a new preference for monism (or at least a monism with multiplicity) had emerged, ascribed to the rather plastic category 'Pythagoreanism,' and

---

[45] There is an important lacuna in Porphyry's account. Note that ποσότης (feminine) is linked not with ταύτην but with τοῦτο (neuter) at line 10, a τοῦτο with no obvious antecedent, but echoed at lines 15–16: τοῦτο τὸ εἶδος (itself asking us to remember an antecedent not readily obvious). If the two τοῦτος are to be linked, then form is a peer or immediate subordinate of quantity.

[46] Whereas μὴ ὂν is at the bottom of Moderatus' Platonist system, posterior or equal to the material world, in the Pythagorean system μὴ ὂν precedes it, with existence and quality as intermediaries. No doubt Moderatus and his sources were grappling with the *Parmenides* and the *Sophist*—the Platonic dialogues that deal chiefly with nonexistence. But with the loss of context, it is hazardous (and for this study tangential) to infer the reason for the difference in the two systems.

[47] Tornau 2001.

[48] For other relevant primary sources and discussion, see Rist 1962, Rist 1965, and Thomassen 2006:270, 275–279, and references.

treated as if it were Pythagoras' original teaching. Pythagoreanism was a literary edifice, a memory to be filled with lore and ideas, not a living community.

The new monism was influential. We see its attraction in Philo of Alexandria, who occasionally used it to describe the divinity, treating God and his powers as if they were part of a Middle Platonist metaphysical system.[49] And, because the new philosophical monism was attractive to monotheists such as Jews and Christians, we will see in the next several chapters how this new monistic metaphysics, combined with the widespread interest in number symbolism, became the foundation for the early Christian theology of arithmetic.

---

[49] Compare *On Flight and Finding* 94–95 (the five powers of the one God) with *On Abraham* 121 (two powers). The two accounts are inspired by numbers Philo encounters in Scripture; hence the disparity. Because Philo molds each account to suit the number, one should not rely upon them as maps to his metaphysics.

3

The Rise of the
Early Christian Theology of Arithmetic

The Valentinians

THE NEW TESTAMENT SHOWS THAT THE EARLIEST CHRISTIANS were attuned to the number symbolism of their day. Christ itemized his followers symbolically, choosing 12 disciples and 70 apostles. The Book of Revelation is adorned with numerous sevens and twelves, and a single, infamous 666. That impulse continued after the apostolic period. In the so-called *Epistle of Barnabas*, an early second-century text, the 318 servants of Abraham (Genesis 14.14) are said to foreshadow Christ: the Greek numeral 318, τιη´, stands for the Cross (from the shape of the τ) and the first two letters of Jesus' name (ιη).[1] The *Shepherd of Hermas*, of roughly the same period, is riddled with number symbolism, its meaning obscure.

Although this interest in number symbolism was widespread, it was neither systematized nor exaggerated. Early Christian writings from the first century, like contemporary Jewish writings, are adorned with but not built upon number symbolism. In the late second century that changed. There emerged new Christian movements that embraced elaborate, mathematically structured theological edifices. They gravitated to numbers that appeared in the Bible, interpreting even the most innocuous ones in light of their theological metaphysics. This new emphasis on number sprouted in a variety of systems, each more complex than the last.

In the next three chapters I describe some of the most striking and best preserved of these early Christian theologies of arithmetic. I discuss the upper echelons of their theological edifices, emphasizing the arithmetical patterns that came into play. I also note the arithmological strategies deployed for inter-

---

[1] "Barnabas" *Epistle* 9.7–9. On second-century Christian and Jewish exegeses of Genesis 14.14 see Ferguson 1990, Hvalvik 1987, Lieberman 1987:168n47, and Gevirtz 1969.

preting the Bible. I describe not only the teachings but the rationale behind them, a rationale that was rooted in classical number symbolism and Pythagorean metaphysics. Although proponents of these systems were part of church life, they were fascinated intellectually and spiritually more with classical number symbolism and Pythagorean-cum-Platonist metaphysics. In their theology they tried to synthesize and challenge religious and secular traditions, to chart new intellectual and spiritual territory, not just in the churches but in late antique culture.

In this chapter I recount the systems of the Valentinians, who were the earliest and most significant part of this movement. Our sources for Valentinianism are of two types: sympathetic and hostile. The hostile sources are, of course, the orthodox heresiologists: early ecclesiastical writers, such as Irenaeus and Hippolytus, whose concern for doctrinal purity determines what and how they quote from the Valentinians. Of sympathetic sources, only a handful of literary fragments attributed to specific Valentinians of uncertain date—Valentinus, Ptolemy, and Heracleon—have been preserved, all culled from hostile sources. All other texts written by Valentinians come from the famous Nag Hammadi collection of texts, and they are notoriously difficult to date, attribute, or even classify.

It is right to be wary of hostile sources. But Valentinianism cannot be reconstructed primarily from the sympathetic ones. Any attempt to determine which Nag Hammadi texts are Valentinian must begin with a typology, and the only typological landmarks we have come from the Fathers. To the same degree that the Fathers have misunderstood the main features of Valentinianism, we too have probably mischaracterized what is Valentinian in the Nag Hammadi library. So to avoid utter despair, we must begin with Irenaeus and the heresiologists for our understanding of Valentinianism and for our classification of the Nag Hammadi material.[2] But we do so critically. We need not—indeed should not—accept at face value the heresiologists' reports, which are marked by inconsistencies and rhetorical exaggeration. As we shall see, however, this is not a critical point for this study, since both hostile and sympathetic sources yield a consistent picture of the Valentinian theology of arithmetic.

---

[2] Desjardins 1986. Five texts are certainly or very probably Valentinian: *The Tripartite Tractate* (Nag Hammadi [NH] 1.5), *The Gospel of Philip* (NH 2.3), *The (First) Apocalypse of James* (NH 5.3), *The Interpretation of Knowledge* (NH 11.1), and *A Valentinian Exposition* (NH 11.2). Two texts are possibly Valentinian: *The Gospel of Truth* (NH 1.3/12.2) and *The Treatise on the Resurrection* (NH 1.4). Thomassen argues for this classification (1995:244). See also Thomassen 1994–1995.

The Origins of the Theology of Arithmetic: Valentinus, Ptolemy, and the Early Valentinians

From the late 130s through the 150s there lived in Rome a certain Valentinus, who was said to have come from Egypt.[3] Through his writing and preaching he attracted a following of Roman Christians, a coterie whose members were soon called Valentinians.[4] What survives of his literary corpus suggests that his teachings, and those of the early Valentinians, were speculative and theoretical. Although Valentinus had little to do with the directions his later followers took, he cultivated in his circle an esteem for philosophically shaped theology.[5]

The *Gospel of Truth*, a homily dating to the mid-second century, and quite possibly written by Valentinus, exemplifies this original Valentinian impulse.[6] It explores the upper levels of the divinity, explains how error entered the world, and relates how the world found salvation. The central theme of the *Gospel of Truth* is a knowledge-based path to restoring people to the incomprehensible, inconceivable Father. To analyze the entire text would take us too far afield.[7] But the metaphysics of the *Gospel of Truth* anticipates the themes that would blossom into the later arithmetical systems. The three primary characters are the Father, the Son/Word, and a special class of beings called 'aeons,' collectively called the Pleroma (πλήρωμα).[8] The aeons search after the Father and, failing

---

[3] The principal texts concerning Valentinus: Irenaeus *Against Heresies* 3.4.3; Epiphanius *Panarion* 31.2.3; Tertullian *Against the Valentinians* 4.1; and Justin Martyr *Dialogue with Trypho* 35.6. See also Thomassen 2006:417–422.

[4] A following is attested in 155 (Justin Martyr *Dialogue with Trypho* 35.6: οἱ δὲ Οὐαλεντινιανοί). The term 'Valentinian' is used by the orthodox heresiologists, not by the Valentinians themselves, at least in the texts that remain. Thomassen regards the term as pejorative, and therefore sets it in quotes (2006:4). But in antiquity one person's slur could be another's honor. As noted below, Irenaeus associated the Valentinians with Pythagoreanism and meant it to be an insult, but they embraced the association. So maybe they called themselves Valentinians, too. In any case the term 'Valentinian' remains useful, and need not be colored by Irenaeus' polemic.

[5] Like Tertullian, scholars were generally skeptical that Valentinus had anything to do with Valentinian doctrines. But Sagnard, who tried to connect the teacher to his tradition, shaped scholars' attitudes in the later twentieth century (Sagnard 1947). See Stead 1980:75–76. For arguments for reverting to the traditional view, see Markschies 1992 and TRE 34:495–500, now tempered by Thomassen 2006, esp. 430–490.

[6] On the date and authorship, still debated, see Thomassen 2006:146–148.

[7] See the insightful summary and analysis in Thomassen 2006, chap. 17.

[8] The specialized term 'aeons,' used throughout Valentinian and gnosticizing literature, applies to emanations from the Father/Monad, sometimes in concert with the Dyad. The identity and function of these aeons are ambiguous. On the one hand, aeons are individuals, participants in the cosmological drama. They have volition, duties, and the capacity to act independently. On the other hand, they are abstractions. Their harmonic interdependence is treated as essential to the well-being of the universe. Their names (see below) give clues as to what they symbolize within the larger philosophical structure. They are also sometimes called powers (δυνάμεις), roots (ῥίζαι), or sources (ἀρχαί). As powers, they act with force and resolve. As roots, they can

to find and know him, fall into the error of ignorance and experience a fog of anguish. We ourselves are those aeons, and are still within the Father, whose desire to be known prompts him to send the Savior to open up to the totality of aeons the knowledge of the Father that is already latent within their hearts. Salvation is portrayed as a purifying transformation of multiplicity into unity, matter into fire, darkness into light, and death into life (Nag Hammadi [NH] 1.3:25.3–19). In this drama there thrive other entities or aspects such as the Holy Spirit, Truth, depth, and spaces. Their roles and relationships are rather amorphous, but overall they support this mission of salvation.

In the cosmology of the *Gospel of Truth*, arithmetic is a subtle and subdued element. The most important arithmetical motif is the transformation of multiplicity into unity, a dominant, guiding principle for the text's one explicit and colorful use of exegetical number symbolism, that of the number one hundred. In discussing the role of the Savior, the text invokes the metaphor of the good shepherd and the parable of the finding of the lost, hundredth sheep. According to the *Gospel of Truth*, the Savior rejoices when he finds it, since ninety-nine is a number "held by" the left hand (NH 1.3:31.35–32.16, Matthew 18.12–14, Luke 15.4–7, *Gospel of Thomas* 107). With the addition of the hundredth, the number passes to the right hand. The lost sheep, then, represents the restoration of a lost unity. The Father is the right hand, which draws in, and therefore perfects, the totality of aeons, represented by the ninety-nine on the left.

The *Gospel of Truth* plays upon a widespread practice, that of the finger calculus. In antiquity, numbers up to ten thousand could be counted on the hands. One through ninety-nine were handled completely on the left (thereby freeing the right hand in the majority of small transactions to do other tasks, such as pay out coins), and the hundreds and thousands were reckoned on the right. So once one reached ninety-nine, which employed all the fingers of the left hand, the next number, one hundred, required one to drop the left hand and use merely the pinkie of the right. Furthermore, to move from left to right was to move from the inauspicious hand to the auspicious. So the *Gospel of Truth* interprets the parable by invoking a combination of symbols—numbers and body parts—to illustrate vividly how multiplicity can and should return to unitary simplicity.[9]

The rather unadorned pleromatology of the *Gospel of Truth*, and its basic motif of the restoration of unity from multiplicity, developed as Valentinus gave

---

    sprout new life. As sources, they form the spatiotemporal foundations for subsequent entities. As aeons (αἰῶνες), they rule over timeless epochs.

9   This interpretation of the Good Shepherd was popular in early Christianity, among both orthodox and heterodox, and may go back to the first century. See Jerome *Letter* 48.2, Bovon 2009:32, and Williams and Williams 1995 and extensive references and studies cited there, to which should be added the finger calculus of Marcus, discussed in the next chapter.

way to the Valentinians. His notion of the Pleroma soon burgeoned into an elaborate, mathematically designed system. According to Tertullian (fl. early third c.), a disciple of Valentinus named Ptolemy was responsible for this innovation.[10] He may well be right. In the *Letter to Flora*, Ptolemy's only extant writing, number symbolism is a recurrent organizing principle.[11] In his cosmology there are three divine agents: the devil, the perfect god, and standing between them an intermediate god who created the universe. Ptolemy claims that the five books of the Mosaic Law, the main subject of the letter, have three separate authors: the intermediate god, Moses, and the elders.[12] This threefold division of the Law mirrors the threefold arrangement of perfect god, intermediate god, and devil.[13] The part of the Law that is authored by the intermediate god is itself tripartitioned into pure, mixed, and symbolic.[14] The Ten Commandments are the most perfect of these three sections, its perfection made evident by the number ten.[15]

Ptolemy, then, shows a fondness for tripartite schemes and the number ten. But the *Letter to Flora*, which concentrates on lower, earthly matters and the origins of the Law, mentions nothing about aeons, powers, or Pleromas—the upper realm of the world. He says that he has reserved that lesson for a future discussion, on how contrary natures, namely the devil and the intermediate god, could come from a single principle.[16] So we do not know how, if at all, his penchant for number symbolism affected his metaphysics, which may have been monistic. But other Valentinian systems close in time to Ptolemy developed new, highly arithmetical theories about the uppermost Pleroma. The earliest, simplest systems reported by Irenaeus cannot be dated, and their development is difficult to trace. But they give a sense of how other followers of Valentinus were inclined to theologize according to mathematical models.

Epiphanes

One of the simplest of the early Valentinian protologies comes from someone Irenaeus describes as a 'renowned' (ἐπιφανής) teacher, a description a later

---

[10] Tertullian *Against the Valentinians* 4.2, 5.1. His knowing the difference between master and disciples explains why Tertullian doesn't call his treatise *Against Valentinus*.

[11] The letter is preserved by Epiphanius *Panarion* 33.3.1–33.7.10. For a convenient English translation see Layton 1987:308–315.

[12] Epiphanius *Panarion* 33.4.1–2, 14.

[13] Epiphanius *Panarion* 33.7.3–4.

[14] Epiphanius *Panarion* 33.4.2, 14; 33.5; 33.6.1–5.

[15] On perfect numbers see n. 78 below. Ptolemy's observation parallels Heracleon's. See pp. 132–133 below.

[16] Epiphanius *Panarion* 33.7.9.

heresiologist converts into a proper name, Epiphanes (*Against Heresies* 1.11.3).[17] To Epiphanes—to adopt his posthumous name for convenience—the primal aeon, Foresource (Προαρχή), exists before all entities and thoughts. This aeon is ineffable and unnamed. Ironically, Epiphanes names it: Monotes (Μονότης, meaning 'unity'). With Foresource/Monotes exists a second aeon, Power (Δύναμις), which Epiphanes names Henotes (Ἑνότης, also meaning 'unity'). The two, being one, produce in turn Monad (Μονάς), also called Source of All (Ἀρχὴ τῶν πάντων), an unbegotten, invisible aeon. Another consubstantial power coexists with Monad, which Epiphanes names One (Ἕν). These four aeons then project all the remaining aeons.

Thus Epiphanes sets at the center of his protology a Tetrad, whose aeons correspond closely in name, function, and structure. Monotes and Henotes are a single thing (ἓν οὖσαι), and they "advance without emitting" (προήκαντο, μὴ προέμεναι) Monad, the third aeon. The paradoxical language is intended to combine in a single image the unity and the multiplicity of the highest aeons. The primal Tetrad is both an individual entity and a quartet of unities—a complex unity. The unities relate to each other in a hierarchy and sequence modeled upon their numeric character. Each aeon's name is some variation on the word 'one.' The names follow a pattern, with the root μον- forming the names of the first and third entities, and ἑν- those of the second and fourth. And when identified individually, the odd aeons are called "sources" and the even "powers" (but as a collection they are called "powers" indiscriminately). The Tetrad can be arranged in a square:

Ἄρχαι	Δυνάμεις
Μονότης	Ἑνότης
Μονάς	Ἕν

Epiphanes' square protology resembles other popular neo-Pythagorean foursomes, for example the four disciplines of the quadrivium, the four elements (fire, air, water, and earth), and the first four numbers. But Epiphanes is not merely indulging in Pythagorean pop culture. With this he also enters a little-known mathematical and philosophical debate of his time, over the differentiation and definition of terms. In the writings of pre-Hellenistic philosophers such

---

[17] Epiphanius *Panarion* 32.5, dependent wholly upon Irenaeus' account (as is Hippolytus' *Refutation of All Heresies* 6.38.2–3). Over these next two chapters the reader may find it helpful to have on hand Irenaeus' *Against Heresies*, Book 1. Text (Latin and Greek fragments): Rousseau et al. 1965–1982. Translations: Unger and Dillon 1992 is preferable to the ANF translation.

as Plato, the noun μονάς 'unity' is not sharply distinguished from the abstraction τὸ ἕν 'the One', an adjectival noun. But some time in the Hellenistic period, when new metaphysical levels were being postulated by neo-Pythagoreans, philosophers began to distinguish the two terms to refer to two different levels of reality. By the second century, many of the neo-Pythagoreans were championing the notion that the monad exists on a plane higher than the One, against the opposite tendency to prioritize the One (see Excursus A). Epiphanes adopts the neo-Pythagorean position, by putting Monotes and Monad prior to Henotes and One. And he doubles the pattern, by suggesting that as the Monad is to the One, so the concept of unity (Monotes and Henotes) is to its instantiation (Monad and Hen). This second distinction replicates the common differentiation between numbers and numerable things (see Excursus A), but places it on the highest metaphysical level. Epiphanes was aiming in his system for philosophical coherence. Perhaps behind Irenaeus' brief report we see a theologian trying to contribute to the philosophy of his day.

The Valentinian Ogdoad

The most common early Valentinian systems are those in which the Pleroma is said to consist of an octet of aeons. Irenaeus discusses many of these systems. A prime example comes from a group who, Irenaeus quips, thought themselves "more prudent" than the Ptolemaeans (*Against Heresies* 1.12.3). According to Irenaeus, these "more prudent" Valentinians taught that from Forefather (Προπάτωρ) and his Thought (Ἔννοια) emanate all at once six other aeons, thus producing the Ogdoad. The emanation follows a certain sequence (see Figure 1). When the Forefather decided to emit something, it was called Father (Πατήρ, the third aeon). And because it was true, it was also named Truth (Ἀλήθεια, the fourth). Then, when the Forefather wanted to show himself, Human (Ἄνθρωπος, the fifth aeon) was generated; and those whom the Forefather preconceived were named Church (Ἐκκλησία, the sixth). Human then spoke the Word (Λόγος, seventh), and Life (Ζωή, eighth) followed on the heels of Word. This completes the upper Ogdoad.

In this system the existence of an original Dyad—Forefather and Thought—is assumed. The Dyad emits the aeons both outside time (all at once) and in a sequence, one that begins with certain activities of the Forefather and terminates in the fruits of otherness. This Valentinian system explains multiplicity as a function of properties, actions, and naming. New entities emerge whenever Forefather or one of the other aeons acts or exhibits a special property. Those acts become objects, and the reification permits the object to be named. Hence Truth comes about only when it can be named as a property of Father.

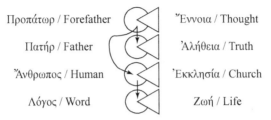

Figure 1. The anonymous "more prudent" Valentinians' system of the aeons, based on Irenaeus *Against Heresies* 1.12.3. Male aeons are designated by triangles; females, by circles. (Illustration by author.)

Names are important in this Ogdoad, just as they are in Epiphanes' Tetrad. They are patterned to reinforce the numerical structure of the aeonic realm. They indicate hierarchy and function. For example, the sequence Forefather → Father → Human → Word follows a descending chain of being, proceeding from that which precedes intelligence and begetting down to that which is the product of intelligence and begetting. The overtones of the names' meanings (especially on the left side of Figure 1) mark how one level relates to the others. The names also mark gender within each dyad, with masculine terms for the aeons in the odd-numbered positions and feminine for the even-numbered ones. Gendered aeons are essential to many Valentinian protologies. In this Ogdoad, each new dyad is generated by a male aeon above it, seemingly without the participation of its female counterpart. So all action and initiative is the responsibility of the male alone, a treatment of gender and procreation that stresses a monadic chain of generation, to complement and offset its dyadic structure.

In other Valentinian Ogdoads the same entities are given different (or multiple) names. They are based on various conceptual schemes, with different lines and sequences of generation. These Ogdoads can be thought of as emerging as a whole, or as one Tetrad begetting another, or as four pairs of dyads emerging seriatim in conjugal harmony. The model was flexible enough to allow various descriptions of the descent. But every Ogdoadic system pictures the result as four tiers of paired entities.[18]

---

[18] The other, simpler Valentinian ogdoads, in order of increasing complexity, are: (1) that of Secundus, who says that there is a right Tetrad and a left Tetrad, corresponding to light and darkness (*Against Heresies* 1.11.2, reported also by Epiphanius [*Panarion* 32.1] and Hippolytus [*Refutation of All Heresies* 6.38.1]); (2) an anonymous system in which the "male" aeons are generated first, then the "female" (the names used are nongendered and highly abstract, focusing on perception; *Against Heresies* 1.11.5, these too are reported by Epiphanius [*Panarion* 32.7] and Hippolytus [*Refutation of All Heresies* 6.38.3–4]); (3) the Barbelo-Gnostics (a group perhaps only influenced by the Valentinians), whose Ogdoad both comes from a preexisting pair (Barbelo and Father) and generates a little ogdoad of its own (among other entities; *Against Heresies* 1.29.1–4, recounted in Theodoret *Compendium of Heretical Fables* 1.13 and various versions of the

A Ptolemaean Protology

Another simple protology comes from "those more knowledgeable around Ptolemy," in Irenaeus' sardonic words (*Against Heresies* 1.12.1).[19] This system is the only one that explicitly fulfills a promise Irenaeus makes in his preface, that he will address, as far as he is able, the notions propagated by the followers of Ptolemy (*Against Heresies* 1.pref.2). It gives a sense of the intellectual trajectory Ptolemy pursued, even if it can be credited only to his followers.

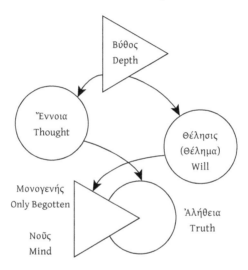

Figure 2. The "more knowledgeable" Ptolemaeans' system of the aeons, based on Irenaeus *Against Heresies* 1.12.1. Male aeons are designated by triangles; females, by circles. (Illustration by author.)

The Ptolemaeans start with a single power, Depth (Βύθος), and claim he has two "dispositions" who are also his consorts: Thought and Will (Θέλησις; see Figure 2).[20] Thought has priority over Will (indeed, Will is said to be Thought's power), but she depends on Will so as to project other powers. The Ptolemaeans explain this relationship by noting that Depth, when he issued them, thought before he willed, therefore Thought has priority over Will. Thought mixes with

---

Apocryphon of John); and (4) systems (discussed below) of thirty aeons, which build upon a core Ogdoad. For a full analysis of the differences, see Thomassen 2006, chap. 20.

[19] Οἱ δὲ περὶ τὸν Πτολεμαῖον ἐμπειρότεροι. Also reported in truncated fashion by Hippolytus: *Refutation of All Heresies* 6.38.5–7. Irenaeus' comment can be read in two ways: either that one Ptolemaean group was smarter than the rest, or that all the Ptolemaeans were smarter than the other Valentinians.

[20] Dispositions: διαθέσεις, a special Ptolemaean term suggesting the arrangement or state of Depth and reserved for the second and third powers.

Will, and out of their conjugal union (συζυγίαν) comes the projection of Only Begotten (Μονογενής, also called Mind, Νοῦς) and Truth. They proceed in this order because, in typical Valentinian fashion, the former (Μονογενής) is male and the latter (Ἀλήθεια) female, as signified by the grammatical gender of their names. Truth is the visible manifestation of Thought, and Only Begotten of Will.

Behind this bewildering account lies a simple pattern, a variation on the standard Valentinian Tetrad. Depth was a common name for the first aeon (the Forefather), and the third aeon was often also called Only Begotten or Mind (not just Father). The Ptolemaeans have added to the Valentinian Tetrad the aeon Will. They have subtly footnoted this modification by calling Will "added later" (ἐπιγενητός), a term that identifies her adventitious role. In contrast, her "sister," Thought, is called "uncreated" (ἀγένητος), a description more befitting the second aeon of the other Valentinian systems.

In their metaphysical edifice the Ptolemaeans use gender to change very subtly the organization, and therefore the geometry, of the divine realm. For example, Will is introduced, in part, to confuse the gender balance and conjugal harmony found in the traditional Valentinian Tetrad. She becomes Depth's second consort.[21] What would seem to lay the foundation for a polygamous, even incestuous, beginning of the aeons mutates even further when the two daughters of Depth enter into conjugal union with each other, to generate Only Begotten and Truth. The older "daughter" Thought, not Depth himself, plays the traditional male part in the seemingly lesbian relationship. And Will herself changes gender. As the consort of Depth, and then of Thought, she is named with a feminine abstract noun, Θέλησις. But this changes in the middle of the story to the neuter cognate Θέλημα (a terminological change also noticed by the ancient Latin translator). Once the male aeon Only Begotten is projected, we learn that Will, not her more masculine sister, is his archetype. Thus, Will starts off female—a subservient one at that—and then transforms into a conduit for maleness. The two female aeons, the "dispositions" of Depth, conjugally unite to generate the first male-female pair, and their relationship develops into a gendered chiasm.

---

[21] Thomassen argues that Irenaeus, not the Ptolemaeans, call Thought and Will consorts (σύζυγοι) of Depth (2000:208–209, arguing on the basis of *Against Heresies* 1.12.2). But this argument assumes that any time Irenaeus associates doctrines of a group with a preceding topic, he invents the connection. This is a possibility, but so is its opposite. Irenaeus claims that the Ptolemaeans "say [Depth] has two consorts, which they also call dispositions." So Irenaeus states that both 'consorts' and 'dispositions' are the Ptolemaeans' words. If we accept that 'dispositions' was a Ptolemaean term (as Thomassen does), why not the other too? Further, the Ptolemaeans build upon Valentinian systems that describe Thought/Silence as Depth's consort. And to introduce two consorts helps the Ptolemaean effort, evident elsewhere, to subvert the gender patterns of most Valentinian protologies.

The Ptolemaeans' gender alterations to the Pleroma disturb the equilibrium of the traditional Valentinian theology of arithmetic. None of the powers is given a numerically derived name, and they are not explicitly grouped into Dyads or Tetrads. Their interrelations no longer follow the arithmetical rhythm of pairs found in standard Valentinian Ogdoads. The total number of entities is five, not the expected four. The chiastic generation of the powers adds crisscrossing elements to the traditional horizontal and vertical movements. This complex geometry justifies the special term given to Thought and Will, 'dispositions,' which implies spatial order as well as personal behavior.

This system shows that various Valentinian groups experimented with and challenged other models. Perhaps these variations were offered in a spirit of competition, perhaps of collaboration. No matter the motives, stable arithmetical patterns in Valentinian theology were not a given.

The Extended Valentinian System

The form of Valentinianism most familiar today espoused thirty aeons, the Triacontad. In this system, which seems to have been widespread and popular even then, the classical Ogdoad is expanded to include a Decad and a Dodecad. Six versions of this Valentinian protology are attested, including one from the Nag Hammadi texts.[22] Irenaeus devotes nearly a third of the opening of his first book of *Against Heresies* (1.1–9) to publicizing and ridiculing this brand of Valentinianism, thereby providing us our most complete account of the Triacontad. In this section I follow his account, but draw from others as needed.

Many scholars refer to this group as 'Ptolemaean,' based on Irenaeus' promise in the preface to discuss the teaching of the followers of Ptolemy. Some scholars also refer to this as the 'grand' or 'main' system, based on its length and prominence in Irenaeus' argument. But these labels slightly overinterpret the text. In the preface, Irenaeus promises to discuss two groups: the followers of Valentinus and those of Ptolemy, the latter merely "insofar as I am able." Thus the system discussed in chapters 1 through 9 could refer to either group, and Irenaeus (uncharacteristically for book 1) never says which.[23] So it would be

---

[22] *A Valentinian Exposition*, Irenaeus *Against Heresies* 1.1–9 and 1.11.1, Tertullian *Against the Valentinians*, Hippolytus *Refutation of All Heresies* 6.29–36, Epiphanius *Panarion* 31.5–6.

[23] Καί, καθὼς δύναμις ἡμῖν, τήν τε γνώμην αὐτῶν τῶν νῦν παραδιδασκόντων, λέγω δὴ τῶν περὶ Πτολεμαῖον, ἀπάνθισμα οὖσαν τῆς Οὐαλεντίνου σχολῆς, συντόμως καὶ σαφῶς ἀπαγγελοῦμεν. Markschies 2000:251; Irenaeus *Against Heresies* 1.pref.2. Irenaeus' promise to treat the Ptolemaeans only as far as he was able, Markschies argues, indicates that he could treat them only briefly, namely at *Against Heresies* 1.12. Thomassen 2006:18 et passim regards *Against Heresies* 1.1–9 as Ptolemaean, but he does not deal with Markschies's arguments, which are in my view more compelling.

hasty to say it is Ptolemaean. And nowhere does he imply that it is the archetype of Valentinianism. True, it is the one Irenaeus discusses at the greatest length, but as we shall see, he considers Marcus to be the apex of Valentinian doctrinal development. The Valentinian Triacontad, for Irenaeus, is a fine specimen of a story that kept bifurcating and changing.[24] Its prominence in *Against Heresies* is explained partly by the original plan of Book 1, at least in its first draft: to broadcast and refute Valentinian doctrines that were not widely known.[25] Irenaeus also gives it prominence because it was extensively described in the handful of Valentinian writings he had secured so that his claims could be corroborated. Thus to call this the 'grand system' reflects our own impressions and the accidents of survival, and so I refer to it here as Irenaeus' 'extended' or 'first' system, or merely as the Triacontad.

The complexity of the ideas and Irenaeus' terseness make this extended system very difficult to understand, even after two or three very slow readings. So here I summarize and streamline the narrative, focusing on the very strong mathematical character of the Triacontad (captured visually in Figure 3).

In this first system the preexistent, transcendent aeon, called Forefather—also called Foresource and Depth—coexists with his consort, Thought—also called Grace and Silence. Depth impregnates Silence, and she brings forth Mind—also called Only Begotten, Father, and Source of All. At the same time Mind is generated, so too is his consort, Truth. Thus, according to Irenaeus, Depth, Silence, Mind, and Truth are the first, original Pythagorean τετρακτύς, what they call "the root of all" (*Against Heresies* 1.1.1). From this first Tetrad emerges a second: Mind begets Word and his consort Life; Word and Life beget Human and Church. Thus is formed the Ogdoad, "root and foundation of all things."

Word and Life, after begetting Human and Church, project another five pairs of aeons, called the Decad, given the names Profound and Copulation (Βύθιος, Μῖξις), Ageless and Union (Ἀγήρατος, Ἕνωσις), Self-engendered and Pleasure (Ἀυτοφυής, Ἡδονή), Immovable and Intercourse (Ἀκίνητος, Σύγκρασις), and Only Begotten and Bliss (Μονογενής, Μακαρία; *Against Heresies* 1.1.2). Human and Church project six pairs of aeons, the Dodecad: Advocate and Faith (Παράκλητος, Πίστις), Paternal and Hope (Πατρικός, Ἐλπίς), Maternal and Love (Μητρικός, Ἀγάπη), Ever-Mindful and Understanding (Ἀείνους, Σύνεσις), Churchly and Blessedness (Ἐκκλησιαστικός, Μακαριότης), and Desired and Wisdom (Θελητός,

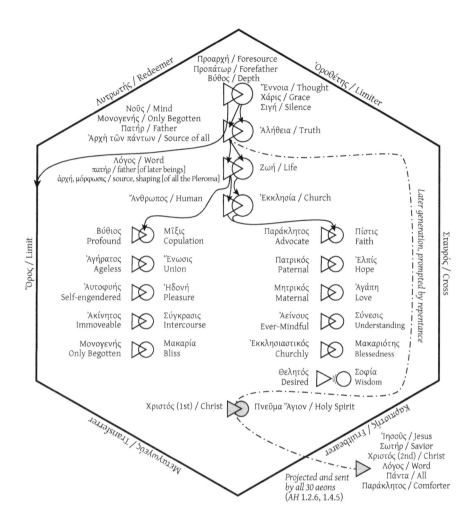

Figure 3. The Valentinian Pleroma recounted in Irenaeus *Against Heresies* 1.1–9. Male aeons are designated by triangles; females, by circles. Hollow triangles and circles represent the base thirty aeons. Arrows indicate lines of projection. The large hexagon represents Limit, who is assigned six names and is said to be hexagonal. (Illustration by author.)

Σοφία). Thus all the emanations combined—the Pleroma—constitute the Triacontad, thirty aeons arranged in three groups: Ogdoad, Decad, and Dodecad (*Against Heresies* 1.1.3; *A Valentinian Exposition* NH 11.2:22–31; see Figure 3).

Sometimes the Triacontad led to yet more aeons. In a variation of this extended system, the Ogdoad, Decad, and Dodecad are synthesized to get 360

(8 × 10 × 12), "the Pleroma of the year," which is then linked to the year of the Lord (Isaiah 61.2).[26] What role these 360 aeons played is unclear.[27]

Like the Valentinian systems discussed above, the aeons' names are thoughtfully planned to augment the mathematical patterns. The name of the first element—the male partner—of every syzygy in the Decad and Dodecad is an adjective of masculine (or common) gender. Female aeons, which make up the even-numbered ranks, are named with feminine abstract nouns.[28] Further, the names of the masculine aeons in the Decad are characteristics of the Forefather; the female aeons' names are terms for sexual intercourse. In the Dodecad, the masculine aeons' names describe the functions of the aeons of the Ogdoad (especially Mind); the female aeons' names describe virtues in a partially recognizable sequence (Faith, Hope, Love: 1 Corinthians 13.13).

The arithmetical patterns of the Pleroma are contained by a geometrical border named Limit ("Ὅρος), a power that the Forefather projects through Only Begotten (*Against Heresies* 1.2.2, 4). The Forefather has created Limit in his own image: it has no consort, and it has no given gender. Limit is also assigned five other names: Cross (Σταυρός), Redeemer (Λυτρωτής), Fruitbearer (Καρπιστής), Limiter ('Ὁροθέτης), and Transferrer (Μεταγωγεύς). The entity is conceived of as a hexagon that delimits and supports the internal Pleroma (*Against Heresies* 1.2.4, 1.3.1). Limit is perceived not only as sixfold but as twofold, principally through its two complementary powers, one stabilizing and one divisive. Insofar as Limit stabilizes, it is the Cross; in its dividing activity it acts as Limit proper (*Against Heresies* 1.3.5).[29]

---

26   *A Valentinian Exposition*, NH 11.2:30.34–38. See also Irenaeus *Against Heresies* 2.22.2.

27   In *A Valentinian Exposition*, the "Root of the All" (i.e. the topmost aeon) goes through three stages of revelation and emanation (NH 11.2:23.32, 26–31), beginning with the 360. It seems puzzling that the 360 would be at the top of the hierarchy. Turner suggests that Mind is the subject, working his way from the bottom of the zodiac, i.e. Silence (1990:154–156). But their interpretation depends upon incorrectly assigning to the monadic Valentinianism of Hippolytus (on which see below) a primary tetrad of Mind–Truth and Word–Life, and it does not explain why Silence and the 360th are to be equated. In other texts from Nag Hammadi the 360 are lower beings: Eugnostos (NH 3.3/5.1) 83.10–20, 84.4–11. Cf. *Gospel of Judas* 49.12–50.3.

28   See *Against the Valentinians* 6.1, where Tertullian complains that the gender of the names cannot be replicated in Latin translation. See also *Against the Valentinians* 11.2.

29   Valentinians differed over the number of powers to assign to Limit. Cf. *A Valentinian Exposition*, where Limit has four powers: separator (ⲟⲩⲣⲉⲥⲡⲱⲣϫ̄), confirmer (ⲟⲩⲣⲉⲥⲧⲁ̄ϫⲣⲟ), form-provider (ⲟⲩⲣⲉⲥ [ϯ ⲙ]ⲟⲣⲫⲏ), and substance producer (ⲟⲩⲣⲉⲥϫⲡⲉⲟⲩⲥⲓⲁ; NH 11.2:26.31, 27.31–33). On this see Turner 1990:99–101, 158–159, and Thomassen 2006:238–240. Thomassen opposes the "they" of 27.34 (assigning to Limit two powers) to the "others" of 27.33 (assigning to Limit four powers), and so treats the Exposition as incorporating two contradictory protologies. But I regard the "they" and "others" as identical. Thus, the "for" (ⲅⲁⲣ) of 27.34 stresses the explanation of, and not the doubt behind, the "why" ([ⲉⲧ]ⲃⲉⲉⲩ) of 27.30. The text at 27.34–38 explains how the first two powers function in Limit, and the last two powers would have been explained on the next

If this unfurling of the Pleroma is act 1 of the Valentinian drama, act 2 is the fall of Wisdom. The first aeon, the Forefather, is unseen and unfathomable, known only to Only Begotten, the third aeon. But Wisdom, the last of the thirty aeons, experiences a desire (in the absence of her male consort, ironically named Desired), a desire that spurs her to seek after and contemplate the Forefather. In her journey into the vastness and inscrutability of the Father, Wisdom is nearly annihilated by the Father's sweetness, and she falls from the company of the Pleroma. She finds herself unable to return (*Against Heresies* 1.2.1–3).

Wisdom strives to be restored to the Pleroma, to be reunited with her original consort. As she repents, she leaves behind certain personal aspects that become entities and characters. She is said to abandon "her former Resolution ('Ενθύμησις), along with the passion that came with that astonishing wonder."[30] In her vain effort to apprehend the Forefather, Wisdom begets a "shapeless essence" (οὐσία ἄμορφος), the kind of nature a woman would beget, were she begetting on her own.[31] Recognizing what has happened, Wisdom experiences first pain, then fear, and finally distress, which leads her to supplicate the Forefather. She is restored, but her Resolution and Passion remain outside Limit (*Against Heresies* 1.2.2–4).

To prevent this tragedy from befalling any other aeon, Only Begotten projects another syzygy through the foresight of the Father: Christ and the Holy Spirit, who fasten and support the Pleroma by teaching the aeons about their syzygies, about the incomprehensibility of the Father, and about thanksgiving and true rest (*Against Heresies* 1.2.5–6).[32] With the restoration of harmony, all the aeons in the Pleroma collectively project an emanation in honor of Depth. This saving emanation has several names: Jesus ('Ιησοῦς), Savior (Σωτήρ), Word (Λόγος), Christ (Χριστός), and All (Πάντα). Later he is called Comforter (Παράκλητος; *Against Heresies* 1.4.5).

---

folio (28), where there is now a large lacuna. The passive verb also fits and reduces the contrast (Timbie's observation).

[30] ἀποθέσθαι τήν προτέραν ἐνθύμησιν, σύν τῷ ἐπιγινομένῳ πάθει ἐκ τοῦ ἐκπλήκτου ἐκείνου θαύματος. Passion (πάθος) may be an aeon too, although Rousseau and Doutreleau do not capitalize it here in their edition. See Irenaeus' critique at 2.20.5 and my discussion below, p. 112.

[31] This alludes to the common belief that in all offspring the male provided the form, the woman the matter. See e.g. Aristotle *On the Generation of Animals* and Plutarch *Isis and Osiris* 53–54 (372e–373a). The notion was common in Pythagorean and Platonist theology of the time. See Thomassen 2006:270–291.

[32] Stead (1969:79) notes that the generation of Christ and the Holy Spirit produce a Pleroma of 32, not 30, aeons, which suggests that Irenaeus is introducing into a dyadic Valentinian system elements of a monadic one. The inconcinnity is noteworthy, but recourse to the monadic/dyadic dichotomy cannot resolve the difficulty. See pp. 52–58 below.

The Valentinian story enters act 3, a discussion of how the Savior enters the fallen world, to make a path of salvation, not just for Wisdom, but for the world she begat. Details of this part of the story are too complex and off-topic for a full summary. But it is worth noting some of the arithmetical patterns, because the world outside the Pleroma is also mathematically arranged.

Just as there are three tiers of beings in the Pleroma (the Ogdoad, Decad, and Dodecad), so the extra-Pleromatic world has three tiers, sparked by the Savior's encounter with Wisdom's Resolution. Through her passion comes matter; through her repentance, the soulish realm; through her pregnancy, the spiritual (*Against Heresies* 1.5.1; see Figure 4). She takes the lessons—μαθήματα, the mathematics, if you will—imparted to her by the Savior, and shapes out of the soulish material an entity called God the Father, also called Mother-Father (Μητροπάτωρ), Fatherless (Ἀπάτωρ), and Demiurge (Δημιουργός). This Demiurge creates soulish things on the right and material things on the left, becoming father to the former, and to the latter, king.[33]

So the hierarchy proceeds from the Pleroma downward, to the spiritual realm of Wisdom's Resolution, to the Demiurge, and finally to the material world. The four tiers of reality are structured numerically, and the entities are given mathematical names. Because the Demiurge creates the seven heavens, he is called the Hebdomad; and Akhamoth (i.e. Wisdom) is called the Ogdoad, thus "preserving the number of the original, first Ogdoad of the Pleroma" (*Against Heresies* 1.5.2–4).[34] As the Ogdoad, she is also the Intermediate Region (μεσότης), bridging the material universe and the Pleroma (*Against Heresies* 1.5.3–4).

The material world is itself tripartitioned, a point repeatedly noted throughout the Valentinian texts.[35] Through the three emotional extremes Resolution experienced while outside the Pleroma—astonishment/perplexity, grief, and fear—come the three elements of the world (not four, as would be expected).

[33] This account parallels the *Tripartite Tractate*, where the fallen aeon (called Word, not Sophia) begets three orders of beings in all. The first two, the psychic and the material, are placed in antithesis, are given opposing epithets, and are said to represent the difference between the Jews and the Greeks (NH 1.5:98.12–20, 109.24–113.1). The third, the spiritual, comes at the advent of the Savior, replicating the joy of the Pleroma and corresponding to humans of a spiritual nature (NH 1.5:93.14–95.38).

[34] ἀποσῴζουσαν τὸν ἀριθμὸν τῆς ἀρχεγόνου καὶ πρώτης τοῦ Πληρώματος Ὀγδοάδος. See also *Against Heresies* 1.3.5, 1.4.1. The same sentiment seems to underlie the Valentinian notion preserved in Clement of Alexandria *Epitomes* 47.1, where Proverbs 9.1 ("Wisdom has built for herself a house and has established seven pillars") is taken to refer to Wisdom and the Demiurge. For other references to the Demiurge as the Hebdomad see *A Valentinian Exposition* (NH 11.2:37.12–15); Irenaeus *Against Heresies* 1.5.2, 1.14.6; Hippolytus *Refutation of All Heresies* 6.32.7.

[35] Cf. *Treatise on the Resurrection* NH 1.4:45.40–46.2 and n. 33 above. In the *Tripartite Tractate*, each tripartitioned part of the material world is assigned a principal vice: arrogance, lust for power, and envy (NH 1.5:103.13–36).

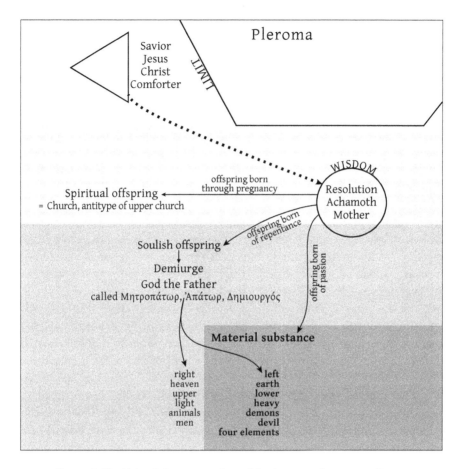

Figure 4. The Valentinian emanation of the lower realms, according to Irenaeus *Against Heresies* 1.1–9. Broken lines indicate activity; solid arrows, generation. The three bands indicate the tripartition of the world outside the Pleroma. (Illustration by author.)

From astonishment/perplexity comes earth; from fear, water; from grief, air. The fourth material element, fire, is said to be latent in all three elements, just as ignorance is latent in all three emotions (*Against Heresies* 1.5.4).[36]

---

[36] Not all Valentinians tinkered with the canonical four elements. In the *Gospel of Philip*, the four elements—earth, water, wind, and light—are matched with the four virtues faith, hope, love, and knowledge (NH 2.3:79.18–30). Here the sequence of the four material elements from dense to rarified provides both a corresponding hierarchy of the virtues and a subtly gnosticizing addendum to Paul, that although love may be greater than faith and hope (1 Corinthians 13.13), knowledge trumps them all. Further, in *A Valentinian Exposition*'s story of Wisdom, the "Tetrad of the world"—presumably referring to the four elements fire, water, earth, and air—is said to "bring

Because of the structure of Wisdom's world, the human is also tripartite, a creature composed of body, soul, and spirit, the three elements derived from earth, the Demiurge, and Akhamoth respectively (*Against Heresies* 1.5.5–6). Reflecting this tripartite structure, human beings, too, are at different levels. Spiritual people—Irenaeus says the Valentinians claimed this category for themselves—are initiated into the mysteries of Akhamoth (*Against Heresies* 1.6.1). Soulish people are the ordinary rank and file in the Church; they lack perfect knowledge and must rely upon good works and bare faith. Earthly people—those outside the Church—have no prospect of salvation. These three moral grades of humanity correspond to Seth, Abel, and Cain respectively (*Against Heresies* 1.6.2, 1.7.5).[37]

According to Irenaeus, in this same system there are those who teach that the Savior descended upon Jesus Christ in the form of a dove, and thereby formed in him a Tetrad, reflecting the first, primal Tetrad in the Pleroma. The Tetrad in Jesus is made up of four aspects: the soulish (provided by Akhamoth), the spiritual (provided by the Demiurge), the dispensation (prepared by an ineffable art), and the Savior (courtesy of the dove who descended on him).[38] Thus Jesus consists of four disjointed natures: spirit, soul, body, and a higher aeonic nature—a sort of superhuman.[39] These natures work independently, so that each Gospel account of Jesus' life was to be interpreted in light of one of his four natures (*Against Heresies* 1.7.2).

This extended Valentinian system may appear outlandish to most readers today, especially since it seems to have little to do with the Bible. But the Valentinians invested considerable energy in establishing the connection. Their exegesis ranges over Old and New Testaments, mining them for numerical references to justify or explain the composition of the Pleroma.[40] The Valentinians

---

forth fruit," in imitation of the Pleroma, or Demiurge (NH 11.2:37.12–15). On the earthly Tetrad see Irenaeus *Against Heresies* 1.17.1, 1.18.1.

[37] In the *Tripartite Tractate*, the three elements spirit, soul, and material are mixed to form the first human, who was accordingly given three (not two: cf. Genesis 2.9) trees to eat from in the garden of paradise (NH 1.5:106.18–31). So all humanity is separated into three categories: spiritual, psychic, or material, in imitation of the Word (i.e. fallen Wisdom), who brought forth these three classes of beings (NH 1.5:118.14–23). For other threesomes in the *Tripartite Tractate* see Attridge and Pagels 1985:23.400–401.

[38] Dispensation: οἰκονομία, a term used by the orthodox to describe the Incarnation of God the Word.

[39] The fourfold model is clarified through Tertullian's interpretation: *Against the Valentinians* 27.2. Irenaeus' "ineffable" dispensation Tertullian takes to be the bodily aspect, and Tertullian turns Irenaeus' ἐκ τοῦ Σωτῆρος into *Sotericiana*, i.e. the (higher, aeonic) nature of the Savior.

[40] Four sections of Book 1 are devoted to Valentinian exegesis (*Against Heresies* 1.1.3, 1.3, 1.8.2–5, and 1.18). The first three come from Irenaeus' first Valentinian group. The fourth, however, comes after Irenaeus' discussion of Marcus, and it summarizes features that apply to Valentinians

show interest not only in direct references—explicitly mentioned numbers—but in more oblique ones, where a sequence or list of terms can be unpacked and numbered to explain the Pleroma. The aeons of the Pleroma, they suggest, were "not spoken of openly because not everyone could comprehend knowledge of them. But they were mentioned mysteriously by the Savior through parables to those so able to understand" (*Against Heresies* 1.3.1).[41] They claim that references to the Pleroma occur throughout the Scriptures and in the Church's liturgical practice. Thus, any instance of 'aeon' in the New Testament or in the Eucharist celebration was to be read as cryptically alluding to the Pleroma of aeons (*Against Heresies* 1.3.1). Paul supposedly taught about the syzygies of aeons in Ephesians 5.32, where, speaking about "the syzygy of life," he says, "This mystery is great, and I am speaking about Christ and the Church." And when Paul uses marriage as a metaphor, he is referring mysteriously to the doctrine of syzygies in the Pleroma (*Against Heresies* 1.8.4).

The prologue of the Gospel of John mentions the upper Ogdoad, according to the Valentinians. They argue that John, intending to discuss the generation of all that exists, at verses 1 and 2 distinguishes God from the Beginning and from the Word (*Against Heresies* 1.8.5). That is, "In the Beginning was the Word and the Word was with God and the Word was God" mentions the first three male aeons, the term 'Beginning' (ἀρχή) here being shorthand for the Source of All (Ἀρχὴ τῶν πάντων), one of Mind's alternate names. The same verses show how one is the projection of the other and how they follow a specific sequence. Verse 4 —"in him was Life"—introduces Life, the consort of Word. A phrase in the same verse, "the life was the light of men," alludes through 'men' to Human and Church, insofar as 'men' (ἀνθρώπων), being plural, must refer to someone other than the singular figure Human. Thus, the entire second Tetrad—Word and Life, Human and Church—is referred to. The first Tetrad is discussed in verse fourteen—"we beheld his glory, glory as of the Only Begotten from the Father, full of Grace and Truth"—where Father, Grace, Only Begotten, and Truth are all mentioned. So John "has clearly indicated" the structure of the two Tetrads in the prologue of his gospel. The lower ogdoad, that of Akhamoth, is represented in the Bible too, most especially in the prophetess Anna, who lived for seven years with her husband, then alone afterward (Luke 2.36–38; *Against Heresies* 1.8.4). Likewise, the Tetrads and the Ogdoad are indicated in Genesis 1.1, where Moses

---

generally. Strictly speaking, the exegesis reported in chapter 18 does not belong to the extended Valentinian system. But it is quite compatible, so I synthesize here all four sections, which share the same patterns of scriptural exegesis. It all likely comes from a single anthology of Valentinian interpretations. See p. 114n20 below.

[41] ταῦτα δὲ φανερῶς μὲν μὴ εἰρῆσθαι διὰ τὸ μὴ πάντας χωρεῖν τὴν γνῶσιν αὐτῶν, μυστηριωδῶς δὲ ὑπὸ τοῦ Σωτῆρος διὰ παραβολῶν μεμηνῦσθαι τοῖς συνιεῖν δυναμένοις οὕτως·

refers to them with the terms 'God,' 'beginning,' 'heaven,' and 'earth' (*Against Heresies* 1.18.1). The second Tetrad, the offspring of the first, Moses refers to soon after with the terms 'abyss,' 'darkness,' 'water,' and 'spirit.' And in honor of the Tetrad, the sun was made on the fourth day, the tabernacle was made of four things, and the stones on the high priest's robe were arranged in four rows (Genesis 1.14–19; Exodus 26.1, 28.17). According to some Valentinians, man was fashioned on the eighth day (not the sixth, as other Valentinians held, in agreement with Genesis), because of the Ogdoad (*Against Heresies* 1.18.2). The Ogdoad is clearly declared also by Noah's ark, which carried eight people, by David's rank as the eighth brother, and by circumcision on the eighth day (Genesis 7.7; 1 Peter 3.20; 1 Samuel 16.10–11; Genesis 17.12; *Against Heresies* 1.18.3).

The Decad is also mentioned in Scripture, namely in the *iota*, the letter used for the numeral 10: at Matthew 5.18, Jesus promises that not one iota will fall away from the law before all is fulfilled (*Against Heresies* 1.3.2). The *iota* as a theological symbol for ten is common throughout early Christianity, as will become evident in later chapters.[42] Here Jesus' very name, ᾽Ιησοῦς, encodes this symbol, denoting the Decad and Ogdoad together in the first two letters, ιη, the Greek numeral for 18 and the standard early Christian *nomen sacrum*.[43] The Decad is proclaimed also in Genesis 1.3–11, in the creation account, where one finds the ten terms 'light,' 'day,' 'night,' 'firmament,' 'evening,' 'morning,' 'dry,' 'sea,' 'plant,' and 'wood,' in that order (*Against Heresies* 1.18.1).[44] Throughout Genesis the Decad is alluded to in numerous passages: the ten nations whose territory God promised to the Hebrews, Sarah's giving her slave to Abraham after ten years, Abraham's servant giving ten golden bracelets to Rebecca, Rebecca's delay for ten days, and the ten sons of Jacob who go to Egypt to buy grain (*Against Heresies* 1.18.3). Elsewhere in the Old Testament, Jeroboam assumes the rule over the ten tribes, the Tabernacle has ten courtyards, and the gates measure ten cubits, all symbols of the Decad. And after the Lord's Resurrection, he reveals himself to the ten disciples (Thomas in absentia) who were hidden, just as the Decad is unseen.[45]

The Dodecad is revealed in Scripture, too: the Lord was twelve years old when he spoke with the teachers of the Law, and he chose twelve apostles (*Against Heresies* 1.3.2, 1.18.4).[46] The mishaps of Wisdom, the twelfth aeon, are

---

[42] See pp. 76, 118, 131 below.

[43] See Hurtado 1998 for the intriguing suggestion that the *nomina sacra* emerged because of the number symbolism latent in the contractions or abbreviations. I owe this reference to Michael Grondin.

[44] See also pp. 128–129 below.

[45] Genesis 15.19–21 (here relying upon the Hebrew: the LXX lists eleven nations), 16.2–3, 24.22, 24.55, 42.3; 1 Kings 11.31; Exodus 26.1, 26.16; John 20.19–24.

[46] Luke 2.42–46; Matthew 10.2; Luke 6.13. See also Clement of Alexandria *Epitomes* 25.2.

alluded to in the apostasy of Judas, in Christ's passion's occurring in the twelfth month of the year of his preaching ministry, and in the woman who had a flow of blood for twelve years (*Against Heresies* 1.3.3). A further example is the twelve-year-old daughter of the head of the synagogue whom the Lord raised from the dead, recalling the salvation of Akhamoth (*Against Heresies* 1.8.2). Just as Moses names the members of the Decad in the creation account, so he does with the Dodecad by referring to 'sun,' 'moon,' 'stars,' 'seasons,' 'years,' 'whales,' 'fish,' 'serpents,' 'birds,' 'quadrupeds,' 'beasts,' and 'man'—twelve terms in all (*Against Heresies* 1.18.1).[47] The twelve sons of Jacob and the twelve tribes also signal the Dodecad (*Against Heresies* 1.18.4). So too the twelve stones on the breastplate and the twelve bells.[48] Moses and Joshua built altars made of twelve stones, twelve men carried the ark of the covenant across the Jordan, and Elisha placed twelve stones around the bull when he contended with the priests of Baal.[49]

The entire Pleroma of thirty aeons is demonstrated by the Savior's not having visibly done anything for thirty years, a vivid symbol of the hidden mystery of the aeons (*Against Heresies* 1.1.3, 1.3.1). The Triacontad is especially evident in the parable of the vineyard, where workers come at the first, third, sixth, ninth, and eleventh hours—1 + 3 + 6 + 9 + 11 is, of course, 30. So too there are the thirty cubits of height in Noah's ark, the thirty elect men among whom Samuel put Saul first, thirty days David hid in the field, thirty who entered the cave with David, and thirty cubits to the height of the holy tabernacle (*Against Heresies* 1.18.4).[50]

For the tripartition of humanity the Valentinians point to the parable of the woman (identified with Wisdom) and the three measures of grain (*Against Heresies* 1.8.3). Paul too, they say, discusses all three classes of human beings.[51]

Throughout their exegesis, Valentinians say the Scriptures "reveal" (μη-νύειν), "make clear" (δηλοῦν), or "point out" (ἐπιδείκνυσθαι) the arithmetic of the Pleroma. Such language might tempt us to think that the Valentinians regarded the Bible as the foundation of their system. But this is to read into the

[47] Matthew 9.20; Luke 8.44, 8.41–42; Genesis 1.14–16, 1.21, 1.20, 1.24, 1.26. Unlike the exegesis of the Decad (see above), this does not follow the order given in Genesis. Also, ἥλιος, σελήνη, and ἰχθύες do not appear in Genesis 1 LXX, and must be inferred. In contrast, the Valentinian decad of creation closely follows the wording of Genesis 1.3–11 LXX.
[48] Genesis 35.22–26, 49.28; Exodus 28.21, 36.21. The twelve bells are not mentioned in the Bible. See, however, Justin Martyr *Dialogue with Trypho* 42.1 and the comments of Rousseau et al. 1965–1982:262.
[49] Exodus 24.4; Joshua 4.9, 4.20, 3.12; 1 Kings 18.31.
[50] Genesis 6.15; 1 Samuel 9.22 (following the Hebrew: LXX has seventy elect); 1 Samuel 20 (but inexactly: three days in both Hebrew and Greek; the Valentinian exegete may have extrapolated thirty from three); 2 Samuel 23.13 (somewhat loosely: three of thirty came to David, in both LXX and Hebrew); Exodus 26.8.
[51] Matthew 13.33; Luke 13.20–21; 1 Corinthians 2.14–15, 15.48.

sources a quasi-Protestantism. To the Valentinians, the Scriptures have hidden meaning that can be unlocked to reveal the numerical structures of the Pleroma. The Bible shows these things "clearly"—the Valentinians frequently use σαφῶς in their exegesis—not because the structures of the cosmos and Pleroma are clear and apparent for all to see, but for the opposite reason. Allusions to or teachings about them in Scripture are hidden and need to be made manifest. That manifestation can be made only to those capable of understanding, namely, to initiates in their inner circle (see e.g. *Against Heresies* 1.pref.2, 1.21.1). To the Valentinians, any knowledge of the aeons of the Pleroma is hidden from everyone in the lower realms—particularly from the Demiurge and his acts of creation (e.g. *Against Heresies* 1.20). Thus the Valentinians claim not that the Scriptures are the source or the basis of their doctrines, but that the doctrines and the knowledge of them explain and unlock the Scriptures.

Mathematics, Metaphor, and Metaphysics

In their protology and their Bible interpretation, the Valentinians developed a number symbolism that was neither haphazard nor capricious. It drew from the mathematics, philosophy, and mythology of their culture, and it was deployed to shape that same cultural terrain. On one level it resembles lowbrow, pop mysticism, but a closer examination shows that the Valentinians used subtle terms and concepts to signal to the cultural elite that they wanted to engage with them in science and philosophy.

For example, in late antiquity there were two principal definitions of number, each quite different. Moderatus of Gades tersely preserves both definitions: "Number is, to speak in a type, [1] a collection of monads, or [2] a progression of multitude starting from a monad and an ascent diminishing in a monad."[52] In the first, collection-of-monads definition, numbers are classes of individual units—for example, a constellation of stars, a pile of pebbles, or some other inert collection. Under this concept a unit is not number proper, but is rather the measure of the set. This is broadly speaking the view Aristotle holds when he says that number is the multitude measured by the unit. Thus, properly speaking, the unit—one—is not a number.[53] This exclusion of one from the set of numerals was

---

[52] Ἔστι δὲ ἀριθμός, ὡς τύπῳ εἰπεῖν, σύστημα μονάδων, ἢ προποδισμὸς πλήθους ἀπὸ μονάδος ἀρχόμενος καὶ ἀναποδισμὸς εἰς μονάδα καταλήγων; following Stobaeus *Eclogae* (Wachsmuth and Hense 1884:1.8). Compare the definitions of Nicomachus, who flourished shortly after Moderatus: "Number is a defined multitude, or a collection of monads, or a flow of quantity composed of monads": ἀριθμός ἐστι πλῆθος ὡρισμένον ἢ μονάδων σύστημα ἢ ποσότητος χύμα ἐκ μονάδων συγκείμενον. *Introduction to Arithmetic* 1.7.1.

[53] Aristotle *Metaphysics* 1088a4–8: σημαίνει γὰρ τὸ ἓν ὅτι μέτρον [5] πλήθους τινός, καὶ ὁ ἀριθμὸς ὅτι πλῆθος μεμετρημένον καὶ πλῆθος μέτρων (διὸ καὶ εὐλόγως οὐκ ἔστι τὸ ἓν ἀριθμός· οὐδὲ γὰρ

widespread in antiquity, even on a popular level.[54] The definition of number as a collection resonated with everyday use, and it was perpetuated in Hellenistic and Byzantine education, thanks in part to Euclid's *Elements*.[55]

By contrast, Moderatus' second, progression-and-ascent definition of number is highly suggestive of movement, transition, change, and cycles. It invokes metaphors of procreation and generation. Number has a start and an end, both termini identical: a single, lofty unit. Plato embraces this definition when he explains how the one and the indefinite dyad generate the numbers.[56] And as we saw in the previous chapter, this notion develops among neo-Pythagoreans to make the unit responsible for the generation of all numbers, indeed of all the universe. The Valentinians, following these Pythagoreans, embrace the second of these definitions to explain the emergence of plurality. The verbs they use, such as 'project' (προβεβλῆσθαι) and 'bring to fruition' (καρποφορεῖσθαι),[57] imply that within the Monad or Dyad there latently reside the aeons it will project, that the 'fruit' (implied by the second verb) was seminally once part of the 'root.' So the Valentinian systems, where entities spawn other entities, and the universe expands and grows outward before it is redeemed and drawn inward and upward, are rooted not only in images of gender and reproduction but in the Platonic definition of number. They are accounts of originate number, and of how reality springs from it. Valentinianism is, in part, a metaphorical expansion of Moderatus' second definition, furnishing it with a vivid mythology; part rationale, part explanation.

Ancient readers acquainted with arithmetic would have recognized the philosophical tenor of Valentinianism through its mathematical patterns. The various levels of the Triacontad are structured along the canonical proportions. For Greek mathematicians such as Euclid and Nicomachus of Gerasa, there were three major proportions or ratios. The first, the arithmetic ratio, defined a number c to be greater than b by the same amount that b was greater than a, for example, the sequences 3-2-1 or 10-7-4. According to the geometric ratio, c was greater than b by the same proportion as b was than a, for example, 4-2-1 or 12-6-3. And in the third ratio, the harmonic, c is greater than a by the

---

τὸ μέτρον μέτρα, ἀλλ᾽ ἀρχὴ καὶ τὸ μέτρον καὶ τὸ ἕν). See also *Metaphysics* N1.1–2, esp. 1087b33–1088a14. See also Annas 1976:36–39.

[54] Burkert 1972:165–166 and Annas 1976:39. The notion can be appreciated in light of certain uses of 'number' in English that preclude singularity, e.g. "she has a number of friends."

[55] Book 7, defs. 1–2: "A monad is that by which each existent thing is called one. Number is a multitude composed of monads": Μονάς ἐστιν, καθ᾽ ἣν ἕκαστον τῶν ὄντων ἓν λέγεται. Ἀριθμὸς δὲ τὸ ἐκ μονάδων συγκείμενον πλῆθος.

[56] Aristotle *Metaphysics* A6 987b29–988a1; N3–4, 1091a13–29; Burkert 1972:22; Annas 1976:43. For a vivid application of this definition of number see Plato *Parmenides* 143c–144a.

[57] Irenaeus *Against Heresies* 1.11.1.

same proportion that *c* minus *b* is greater than *b* minus *a*, for example, 6–4–3 or 15–12–10.[58] The Triacontad mirrors these mathematical principles. The most fundamental series, Ogdoad—Decad—Dodecad, is an arithmetic progression. The emanation of aeons within the Ogdoad, (1)–1–2–4, follows the geometric progression. And the six-sided Limit forms with the Ogdoad and Dodecad a harmonic progression.

The various Valentinian systems also draw deeply from classical number symbolism and lore, some of which can be traced to the early Pythagoreans. Their assignment of odd-numbered aeons to males and even-numbered to females depends upon very ancient symbolic associations of gender with odd and even.[59] The Valentinian habit of giving their aeons names that are symbolically linked to their mathematical function resembles the ancient practice, attested in Philolaus, Xenocrates, and others, of assigning to the numbers one through ten various gods and goddesses, based partly on the criterion of gendered number.[60] In one of the simpler Valentinian systems, the Ogdoad is composed of two Tetrads that unfold from left to right, like diptychs. This Ogdoad is associated with the Pythagorean table of opposites, a list of ten fundamental opposites, made famous by Aristotle, that follow the same left-to-right pattern.[61]

The Valentinians embraced Pythagorean symbolism, both old and new. Irenaeus, when discussing the Valentinian fondness for the number four, calls their first Tetrad "the first and original Pythagorean τετρακτύς, which they also call 'the root of all.'"[62] The redundant term 'Pythagorean τετρακτύς' (see Excursus B) could be interpreted equally well as both Irenaeus' sarcasm and

---

[58] Expressed as formulas, the three progressions are $x, x + n, x + 2n$; $x, nx, n2x$; and $x, 2xy/(x + y), y$ (alternatively: $x, x + n, x[x + n]/[x - n]$).

[59] Numerous sources from classical through late antiquity attest to the association of odd and even with male and female. See references at Burkert 1972:34, 268, 429, 468–469, 475–476.

[60] In Philolaus, fragment 20a, the dyad is called the consort of Kronos. See also his testimony 14, where the angle of a triangle is assigned to male gods, and the right angle is assigned to goddesses. See Huffman 1993:351–352, and Plutarch *Isis and Osiris* 30; Proclus *Commentary on the First Book of Euclid's "Elements"* 130.8–14, 166.25–167.14, 173.11–13, 174.12–14; Damascius *Commentary on "Parmenides"* 2.127.7–17. Xenocrates, fragment 15 (Isnarde Parente 1982:213), calls the monad and dyad gods. This fragment derives from Aetius *Placita* 304/Stobaeus *Eclogae* 1.1.29b.44–57. Note that the monad is Kronos in Philolaus, but Zeus in Xenocrates; common to both is that mythology, not etymology, is applied to odd and even numbers. Compare the later tradition in which Pythagoras is said to have likened the monad to Apollo and the dyad to Artemis. Moderatus, fragment 3; Plutarch *Isis and Osiris* 10.

[61] Irenaeus *Against Heresies* 1.11.2, concerning a certain Secundus, who applies three of the ten opposites: male/female, left/right, light/dark.

[62] Καὶ εἶναι ταύτην πρώτην καὶ ἀρχέγονον Πυθαγορικὴν Τετρακτύν, ἣν καὶ ῥίζαν τῶν πάντων καλοῦσιν· (*Against Heresies* 1.1.1). Cf. Tertullian *Against the Valentinians* 6.6, where the first Valentinian tetrad is called *quadriga*, a four-horse chariot, not a τετρακτύς. Thus the term 'root' need not be interpreted exclusively as a Pythagorean symbol.

the Valentinians' preferred term. Irenaeus holds that theology should not be beholden to any philosophy other than the rule of faith given to the Church, and he presumes that his audience shares that intuition. On the other hand, the Valentinian term 'root of all,' one of their favorite epithets for the Tetrad, invokes the τετρακτύς, described in one of the oldest and most famous Pythagorean texts, the *Golden Poem*, as "possessing roots of everlasting nature."[63] So the Valentinians made the τετρακτύς, like so many other Pythagorean symbols, their own. What to Irenaeus was an embarrassment, an insult, was unapologetically embraced by his opponents.[64]

Number symbolism could also be used to augment their myth of salvation. This is especially evident in Irenaeus' extended system. There the number four is applied to the upper realm, but only infrequently to the lower one, and always as a symbol of intervention and salvation. In the one counterexample, the Savior-like figure called the Lord is said to take on a fourfold character, in imitation of the Tetrad. But this makes him an oddity in a world where the nature of humans is threefold. In this adaptation of the Platonic tradition, which taught that the soul was tripartite, the trifold human nature is a microcosm of the fallen universe.[65] And that fallen realm is everywhere divided in threes. Three kinds of offspring emerge from Wisdom's Resolution-Akhamoth, and they result in spiritual, soulish, and material substance. This tripartition leads to humanity's falling into three classes: the elect, the Church, and those without hope of salvation. The boundaries between these three classes of people were well defined liturgically, through baptism and a special sacrament the Valentinians called redemption (*Against Heresies* 1.21.2).[66] The three human categories correspond to Seth (spiritual and receptive of the seed of salvation), Abel (soulish and not receptive of that seed), and Cain (earthly, material, and wicked; *Against Heresies* 1.5.1, 1.6.1, 1.7.5). Even the material world, which in the ancient world was normally divided into four elements, is tripartitioned, with fire intermixed with water, air, and earth (*Against Heresies* 1.5.4).[67] The three elements are the direct result of the three passions of the Mother Akhamoth: fear, pain, and perplexity. Threes belong to the Fall.[68]

---

[63] See p. 183 below.

[64] The Valentinians used the term τετρακτύς in other contexts, particularly concerning the creation of the world (Irenaeus *Against Heresies* 1.18.1). See also chapter 4 below.

[65] See Stead 1980:92–94.

[66] On this so-called redemption see Thomassen 2006:360–364, 401–402.

[67] See also Clement of Alexandria *Epitomes* 48.4.

[68] See also Clement of Alexandria *Epitomes* 28 and Ptolemy's *Letter to Flora*, discussed above. This principle is not followed in every Valentinian system. The lower world could be organized by fours (Hippolytus *Refutation of All Heresies* 6.32–34 and Irenaeus *Against Heresies* 1.18.1, discussed below), and the upper world by threes (Irenaeus *Against Heresies* 2.15.2; *Tripartite Tractate* passim; Clement of Alexandria *Epitomes* 80.3).

But other Valentinians, not so highly principled in using number symbolism to mark the two halves of the universe, appealed wherever possible to the order of the material world to corroborate their notions of the Pleroma, even if it brought the lower and upper realms together. According to *Against Heresies* 1.18.1, some Valentinians said that the Tetrad, Ogdoad, Decad, and Dodecad are enshrined in the human body: the four senses (touch is omitted), the eight receptacles of those senses, the ten fingers, and the twelve innards.[69] Others argued that the four elements—fire, water, earth, and air—are all projected, and so are an image of the first Tetrad. Along with a derivative, second Tetrad—the energies of these elements: heat and cold, dry and wet—the four natural elements encode the Ogdoad (*Against Heresies* 1.17.1). To these same Valentinians the Decad is indicated by seven circular bodies, an eighth heaven encircling them, and the sun and moon.[70] The zodiac indicates the Dodecad.[71] And since the highest heaven is "yoked against" the orbit of the totalities, it goes from one portion of a sign to the next every thirty years.[72] The motion of the heavens is an icon of Limit, which encloses the Triacontad. Other Valentinian parallels to the natural world are numerous: the moon's circuit is thirty days; that of the sun, twelve months; the day is divided into twelve hours; each twelfth part of a full day is further divided into thirty parts; and the earth is divided into twelve zones.

The Valentinians' appeal to the natural world and Scripture as repositories of the symbols of the Pleroma resembles similar comparisons by Philo and Plutarch, who identify in texts and the material world patterns that reciprocate their favored number symbols. But the Valentinians moved well beyond this occasional use, and pioneered a single, extended cosmology, a symbolic center of the universe that could bring mathematics, science, and Pythagorean lore within its orbit.

Monism versus Dualism

Valentinian theology pushed the boundaries not only of Pythagorean number symbolism but of middle Platonist and Pythagorean metaphysics. Valentinians

---

[69] Eight receptacles: two nostrils, two eyes, two ears, and the division in the tongue between bitter and sweet. The twelve innards are not specified. Cf. Marcus' body of Truth, discussed in chapter 4 and p. 131n20 below.

[70] But in antiquity the sun and moon were included among the seven planets, so this list has counted them doubly. Possibly "sun and moon" here is a euphemism for other celestial entities. Or perhaps this is a muddled attempt to imitate the Pythagorean cosmology of ten spheres (see Burkert 1972:337–347). Our source does not explain.

[71] See also Clement of Alexandria *Epitomes* 71, which reproduces notes by a Valentinian who relates the twelve zodiacal signs to circumstances in life.

[72] This seems to refer to precession (see p. 139n64 below), since it posits a one-degree movement of Earth's axis against the stars every thirty years, more than twice its actual speed.

took up the two major philosophical positions—both traditional Platonist dualism and the newly emergent Pythagorean monism. They took sides, or attempted new syntheses and explanations. Although they were not philosophers, the Valentinians interacted with the philosophical models of their day, and they advanced perspectives that proved to be influential, if only in method.

The story picks up with Hippolytus, an orthodox writer who flourished in Rome in the early third century.[73] His *Refutation of All Heresies*, a ten-book exposé of pagan error and Christian heresy, depends considerably upon Irenaeus' *Against Heresies*, but it includes systems nowhere else attested. In the case of Valentinianism, Hippolytus reports a significantly different variation of Irenaeus' extended system.[74] Hippolytus claims there was a major disagreement between dualist and monist camps, between Valentinians who taught that the Monad has a consort and those who denied it. But he presents details of only the monists, who emphasized the solitary uniqueness of the first aeon, with no primal, conjugal Dyad (*Refutation of All Heresies* 6.29.3).

In Hippolytus' Valentinian system, the Pleroma has but one source, a Monad in every respect. It has no consort, and remains completely independent of all other entities (*Refutation of All Heresies* 6.29.5). This being, called the Father, did not love solitude, and he decided to beget and bring out from within himself the loveliest, most complete thing he could. The Monad begat a Dyad, and this Dyad, Mind and Truth, became the mother and source of all other aeons that are enumerated inside the Pleroma. In imitation of the Father, Mind projects Word and Life (*Refutation of All Heresies* 6.29.6–7).[75] Word and Life, in turn, project Human and Church. Thus the first group of aeons consists of the monadic Father, standing over three pairs of emanations, or powers. The protology of one followed by six, in sum a hebdomad rather than an ogdoad, is a recurrent structure in other groups mentioned only in the *Refutation of All Heresies*.[76]

The rest of the Pleroma resembles that of the traditional Triacontad, but the Pythagorean rationale is made more explicit. Mind and Truth, seeing their offspring productive, present to the Father a perfect number of aeons: ten.[77]

---

[73] Scholars have recently tried to settle once and for all questions as to how many Hippolytuses there were, who they were, which Hippolytus wrote what work ascribed to the single Hippolytus, and so forth, but no consensus has been reached. See the issue of *St. Vladimir's Theological Quarterly* devoted to the *status quaestionis* in 2004 (volume 48, numbers 2–3). My analysis relies exclusively upon the author of the *Refutation of All Heresies*, whatever Hippolytus he may have been.

[74] For a fuller comparison of Irenaeus' and Hippolytus' accounts of Valentinianism, see Stead 1969.

[75] Paralleling Irenaeus *Against Heresies* 1.1.1, even down to a very minor detail: no mention is made of Truth's role in projecting Word and Life.

[76] See e.g. the system of Monoïmus, discussed in chapter 5 below. Stead 1969 argues that such a hebdomadic structure would have appealed to a second-century reader of Philo.

[77] Cf. Irenaeus *Against Heresies* 1.1.2, where Word and Life, not Mind and Truth, beget the Decad, a difference Hippolytus notes at *Refutation of All Heresies* 6.30.4.

According to Hippolytus' Valentinians, the Father had to be glorified by a perfect number since he, the unbegotten Monad, is himself the most perfect. Thus, just as the Monad in its utmost perfection is the foremost of numbers in the Decad, so the Decad is the foremost of things that come into being in multitude (*Refutation of All Heresies* 6.29.8).[78] Word and Life, seeing Mind and Truth glorify the Father, attempt to glorify their own parents. Because they lack the same level of paternal protection, Word and Life beget twelve aeons, a slightly less perfect number (*Refutation of All Heresies* 6.30.1–2).[79] Thus there are in total twenty-eight aeons, not counting the Father, who transcends them (*Refutation of All Heresies* 6.30.3).

Hippolytus' version of the myth of Wisdom's fall is similar in many respects to that of Irenaeus' first system. She is the twelfth (or twenty-eighth) aeon, and she tries to imitate his monadic, syzygy-less state, and does so without properly understanding the difference between the Father's transcendent nature and her own, inferior nature (*Refutation of All Heresies* 6.30.6–7). Wisdom's fall prompts the Father to order Mind and Truth to project Christ and the Holy Spirit, to stabilize the Pleroma and to bring about the full number of aeons, thirty (*Refutation of All Heresies* 6.31.1–3). Just as in Irenaeus' extended system, the exiled aeon Wisdom is called Ogdoad, and the Demiurge, Hebdomad.[80] But in Hippolytus' system there is no upper Ogdoad, so the significance of her epithet is lost, an indication that it postdates Irenaeus' extended system.

---

[78] Cf. *Theology of Arithmetic* 81.9. In ancient number symbolism various numbers are called perfect, most notably 3, 6, 7, and 10, each for a different reason: 3 has beginning, middle, and end; 6 is the sum of its factors (including 1, but excluding itself); 7 has cosmological and theological perfection, especially in the Jewish and Christian traditions; and 10 is the image of 1, the most perfect number. For other ancient discussions of 10 as perfect, see Aristotle *Metaphysics* 986a8, *Problemata* 910b31, and fragment 203 (= Alexander of Aphrodisias *Commentary on Aristotle's "Metaphysics,"* Hayduck 1891:40); Plutarch *The E at Delphi* 9 (388e); anonymous [*On the Numbers*] (Delatte, ed., lines 20, 55); Philo *Questions and Answers on Genesis* 4.110; Clement of Alexandria *Stromateis* 6.11.84.5 (discussed below, p. 131); Monoïmus in Hippolytus *Refutation of All Heresies* 6.24.1–2, 8.14.6 (discussed below, p. 91); Hippolytus *Refutation of All Heresies* 1.2.8–9, 4.51.6, 6.23.5; Porphyry *Life of Pythagoras* 52; Chalcidius *Commentary on the "Timaeus"* 84.5–8; and anonymous, *The Mysteries of the Letters* 11 (Bandt 2007:122).

[79] Cf. Irenaeus *Against Heresies* 1.1.2, where Human and Church beget the Dodecad, a difference Hippolytus notes at *Refutation of All Heresies* 6.30.5. The comment on the relative imperfection of the Dodecad must be Hippolytus', since there is no indication in any other Valentinian system that the number twelve symbolized deficiency. Very rarely were numbers treated as inauspicious, prognostic texts aside. Hippolytus' other heresies are frequently obsessed with the perfection of the number ten (see previous n. and below, chapters 7 and 8), and it is likely that Hippolytus was so struck by the consistency of this motif across different systems that he inferred that the Valentinian Decad must signify perfection, and so conversely the Dodecad imperfection.

[80] Ogdoad: *Refutation of All Heresies* 6.31.7, 6.32.9, 6.33.1, 6.34.8, 6.35.4, 6.36.1. Hebdomad: 6.32.7, 6.36.1, 6.33.1.

Thus Hippolytus' Valentinians exhibit two critical differences from Irenaeus' extended Valentinian system: First, the Monad has no consort: the Dyad emerges separately from and subordinate to the Monad. Second, Wisdom's fall occurs not because she desires to know the Monad, but because she tries to replicate his solitude. She tries, essentially, to become a Monad. Both these differences are central to Hippolytus' contention that there were two kinds of Valentinians: monists and dualists.

But this contention was a gross simplification, and being a careful reader of Irenaeus, Hippolytus should have known so. At *Against Heresies* 1.11.5, after discussing an anonymous Valentinian system, Irenaeus outlines three types of monadic or dyadic Valentinianism, revolving around the original state of the first aeon, Depth: (1) Depth is without consort since he is neither male nor female, nor even altogether subject to existence; (2) Depth is androgynous, encompassing in himself the nature of a hermaphrodite; (3) Depth has Silence as a bedfellow (συνευνέτις), and the two constitute the first syzygy. The three positions are incompatible, and fall along a spectrum ranging from monadic to dyadic. The first position, which claims for the Monad utter solitude, resembles Hippolytus' system. Irenaeus' extended Valentinian system, which emphasizes the paratactic relationship between Depth and Silence, falls in the third, dyadic position. The second position is a middle ground, envisioning Depth as bi-sexual, its own consort. It promotes a vision of the Monad as being intrinsically dyadic. To these Irenaeus adds a fourth position when he moves on to the Ptolemaeans (discussed above), who taught that Depth has two consorts. Overall, Irenaeus is sensitive to the Valentinians' wide variety of protological models, which he takes as evidence of their incoherence and inconsistency.

Hippolytus, however, simplifies the varieties. He reduces the options to two, Irenaeus' first and third positions. To the group that champions asexual monism Hippolytus assigns an origin in Pythagoreanism. He claims that the dyadic group, which he otherwise ignores, was an offshoot, trying to answer a problem faced by the first group, of how generation can come from only a Father (*Refutation of All Heresies* 6.29.3, repeated at 6.38.5). His oversimplification is intentional. By omitting Irenaeus' second, middle position; by discounting the third, dyadic group; and by excluding the fourth, Ptolemaic position, Hippolytus molds Valentinianism to serve his overall thesis, that each heretic stole his error from an earlier philosopher. He sets out to make the Valentinians followers of Pythagoras, so he isolates the monadic strain.[81] One might think that Hippolytus should have emphasized the dyadic school of Valentinians, given the early

---

[81] On such polemics by Hippolytus, see Marcovich 1986:35–38 and Mansfeld 1992.

Pythagoreans' preference for polarities.[82] But the mistake Eudorus had made, that Pythagoras and the earliest Pythagoreans were monistic and that dualism had been introduced later, was pervasive. By Hippolytus' time the older, dualist form had been forgotten.

Unfortunately, many modern presentations of Valentinianism, despite occasional dissent, follow Hippolytus' oversimplification, dividing the school and its texts into only two camps, monadic versus dyadic.[83] To be sensitive to the complexity of the positions is essential for understanding how the Valentinian treatises intended to interact with the philosophy of their day. Take the *Tripartite Tractate*, written probably in the late third century, but based on earlier material.[84] This metaphysical treatise describes the origins of the universe, presented from the top down, and unfolds its story in three stages: the origin of the Father, Son, and aeonic realm; the creation of the material world; and its redemption. The system begins with a preexistent, single Father, unbegotten, uncreated, unchangeable, and immovable. He has neither consort nor coworker, and he creates or begets without any primordial forms, separate material, or internal substance—those who hold to such views are ignorant (a swipe against Platonists and other Valentinians; NH 1.5:53.21–39). With him is the firstborn, only begotten Son, existent from the beginning. The Church—called the "aeon of the aeons"—comes forth from the Father and the Son like kisses, and from there come the aeons, who reside within the Father like a seed or fetus. The Pleroma of aeons undergoes a process of individualization through glorification, and they settle into three levels of glory in the Pleroma (NH 1.5:69). Although the Pleroma is tripartitioned, there is no suggestion of any further organization into Dyads, Tetrads, and so forth. There are, in fact, neither gendered pairs nor genders. Gendered differentiation is the consequence of the fall of the errant aeon, femininity being a sickness and a deviation from the norms of masculinity (NH 1.5:78.11–13, 94.17–18).

So the Pleroma is rather simple, consisting merely of Father, Son, and Church, the last two eternal and latent within the Father. This may suggest that this form of Valentinianism is quite primitive, like that found in the *Gospel*

---

[82] See inter alia Burkert 1972, Staab 2009.

[83] See Stead 1969:77n2–3. The monadic-dyadic dichotomy informs Turner's edition of *A Valentinian Exposition* (91, 97–99) and the commentary of Attridge and Pagels on the *Tripartite Tractate* (1985:22.179–180; 23.218–219). But Attridge and MacRae, commenting on *The Gospel of Truth*, note that in Valentinian systems a primordial principle may also be thought of as dyadic: "It is, in fact, likely that the divergences within the Valentinian tradition on this subject are more matters of emphasis in articulating a complex fundamental theology than they are radically distinct theological positions" (1985:22.77).

[84] On the date see Thomassen 2006:263–266, 1994–1995:302–303, and 1989:18–20; Attridge and Pagels 1985:22.178. I follow the Coptic text in Thomassen's edition.

of Truth. But arithmetic and numerical unity are essential to the description of relationship between the monadic figure (the Father) and the rest of the Pleroma. In the preface, the Father is at first said to be "like a number" (ⲉϥⲟⲙ̄ⲡⲣⲏⲧⲉⲛ̄ⲛⲟⲩⲏⲡⲉ; NH 1.5:51.9–10), but is then immediately said to be unlike a "one" or "solitary individual" (ⲉϥⲟ ⲙ̄ⲡⲣⲏⲧⲉⲛ̄ⲟⲩⲉⲉⲓⲟⲩⲁⲉⲉⲧϥ̄ ⲉⲛ'; NH 1.5:51.11–12).[85] The apparent contradiction is resolved in the role of the Son, whose eternal presence with the Father makes it impossible to speak of the Father only as one. Nevertheless, the Father is singular. The Father's unity is always shared with the Son, who preexists eternally with the Father (NH 1.5:51.12–15, 16, 24; 57.33–59.1). The Son is to the Father "the form of the formless, the body of the bodiless, the face of the invisible, the word of [the] unutterable, the mind of the inconceivable" (NH 1.5:66.13–16; trans. Attridge and Mueller). The Son projects from the Father, "the one who stretches himself out," the term "stretching" being a common philosophical term to describe the monad's departing from itself to become a dyad (NH 1.5:56.2–3, 16–17; 65.4–5; 66.6–7).[86]

The *Tripartite Tractate* is monadic, but it does not exhibit the pure monism found in Hippolytus' system, which emphasizes the absolute solitude and monarchy of the Monad. Although in the *Tripartite Tractate* the Father also has no consort, he is perpetually filled with his offspring, both the Son and the aeons that compose the Church, earning him the epithet 'the Entireties' (πληρώματα). The Father is an incomprehensible plenitude of goodness. The Son and the Church exist from eternity in the thoughts of the Father, which are externalized, like sprouting seeds. The process is not one from nonexistence to existence, as would be expected in a purely monadic system, but of interior latency to external individualization. The *Tripartite Tractate* avoids metaphors of gender, marriage, and procreation, and restricts itself to an eternal hierarchical Father–Son relationship. It thus offers an unusual type of monism, one that makes plurality an essential, eternal condition of the Monad, something similar to Irenaeus' second of three positions regarding Depth.

Sometimes a monistic universe could have pronounced elements not just of pluralism but of dualism. An example is the system of the Barbelo-Gnostics, a group related to the Valentinians (*Against Heresies* 1.29.1–4).[87] According to

---

[85]   Technically there is a problem here, since in antiquity one was not a number (see above) but the source of number: ἀριθμός implies plurality. To solve it, Thomassen suggested that this means "multitude" (πλῆθος), not "number" (ἀριθμός) (1989:261–262), but the problem still remains, since one is also never multitude, which is, anyway, a species of number. See Nicomachus of Gerasa *Introduction to Arithmetic* 1.3.1–2.

[86]   Thomassen 2006:275–277.

[87]   The exact relationship is unclear. See Markschies 2003:94–97.

Irenaeus and Theodoret, they held that the uppermost aeon, the unnameable Father, dwelt within the second one, the virginal spirit called Barbelo.[88] But neither apologist further explains the relationship between the two uppermost aeons. Their testimony on its own would imply a dyadic system. But the *Apocryphon of John*, a Barbelo-Gnostic text that survives independent of the apologists, with its very similar account of the aeonic realm, shows that the system begins with the absolute solitary monarchy of the Monad, frequently called the invisible Spirit. Two pages are devoted to describing its transcendent properties (NH 2.1:2.26–4.26). All things exist in him, and he in nothing else. He is radiant, perfect light, and in that light appears a reflection that becomes the first power, Barbelo. She is his first thought and becomes the womb for all other aeons. But she emerges at a specific stage in the solitary life of the Monad, the invisible Spirit. Although called Father and given the pronoun "he," the invisible Spirit is not otherwise described in terms of gender. He is called virginal, not because he abstains from conjugal relations (Barbelo conceives by him at NH 2.1:6.12–13), but most likely because he transcends gender, just as he transcends all his other predicates. By contrast, Barbelo is androgynous, is called "Mother-Father" and "Thrice-Male," despite her feminine persona and pronoun. So even though Barbelo and the invisible Spirit form a kind of pair, the conjugal aspect is downplayed. One is ungendered and the other bi-sexual. At its core the system is monadic, but because the myth is intended to explain the emergence of multiplicity, it retains a strong dyadic flavor, evident especially in the reports by Irenaeus and Theodoret.

The converse construction was also possible, as some Valentinian systems that were primarily dyadic had a monadic dimension. In his summary of the system of pseudo-Valentinus (*Against Heresies* 1.11.1), Irenaeus presents what seems to be an obviously dyadic version of the Triacontad. Yet that same system introduces two Limits, one encompassing the Pleroma, the other separating the ineffable one, Depth, from his consort and all other aeons. That is, by distinguishing the ungenerated aeon from all other, generated aeons, the first limit preserves a monadic quality in the Pleroma. Another Nag Hammadi text, *A Valentinian Exposition*, also offers a form of dyadic Valentinianism with monadic elements. Previous commentators and editors have treated the text as being primarily monadic, but there are no good reasons to do so (see Excursus C). In fact, *A Valentinian Exposition* describes a Pleroma that is nearly identical to the Triacontad of Irenaeus' extended system. The Father has a companion, Silence,

---

[88] Systems where the second principle envelops the first appear frequently in Pythagorean texts of the period. See e.g. *Theology of Arithmetic* 1.10–12, 3.1–5 and other examples at Thomassen 2006: 293–294.

and they form a dyad. Yet the Father exists monadically, a special solitude he enjoys within his dyadic condition. He dwells in the Monad, Dyad, and Tetrad, and this Tetrad generates the next Tetrad to produce the Ogdoad, although it is not explicitly named so in the extant text (NH 11.2:25.19–20). The third and fourth syzygies—Word and Life, and Human and Church—generate the Decad and Dodecad, respectively (NH 11.2:30.16–19), thereby generating the Triacontad.

Other Valentinian systems reported by Irenaeus are discussed so tersely that it is difficult to classify them in one of his four positions.[89] Tertullian accuses the Valentinians of vacillating between monism and dualism. He snipes at them for introducing a second person to an entity they want to be solitary, both "in him and with him."[90] But this much is clear: the Valentinians offered a variety of mythological models to account for the origins of the universe. One might be tempted to say that the dichotomy has been false all along. But of all the systems I have described, some embrace a sophisticated complexity, whereas others cling to one ideal or the other, avoiding language that would seem tainted by the other side. That efforts to achieve philosophical purity were not always successful should not seduce us into thinking there was no such idealism.

Of the two ideals, monism, a gift of the neo-Pythagoreans, exercised a powerful (but not irresistible) attraction on the Valentinians. The latter, like the former, as Einar Thomassen has noted, employ technical terms of "extension" and "spreading out" to describe theories of how the Monad emerges into multiplicity.[91] For both groups this indefinite expansion is typical of a dyadic entity. The Monad, on the other hand, is characterized by limit and definition.[92] The parallels between the two groups concerning the uppermost unity and multiplicity extend also to the lower world. Sophia in Irenaeus' extended Valentinian system commits an act of audacity that results in separation, otherness, alienation, and the engendering of the corporeal world, a drama that coincides with both of Moderatus' systems.[93] That audacity occurs well after the emergence

---

[89] Although pseudo-Valentinus (*Against Heresies* 1.11.1) seems dyadic, the first aeon, Depth, is separated from the rest of the Pleroma, even from its conjugal aeon; without an original source the ambiguity cannot be resolved. Secundus (*Against Heresies* 1.11.2) does not discuss Depth, and so we are left to suppose, from his emphasis on a primordial Ogdoad divided into left and right Tetrads, that he was dyadic. But this is conjecture. Other systems: Epiphanes is ambiguous (*Against Heresies* 1.11.3); the anonymous Valentinians seem monadic (*Against Heresies* 1.11.5); and the anonymous, "more prudent" Valentinians appear dyadic (*Against Heresies* 1.12.3). But these are once again impressions from the little that is extant.

[90] Tertullian *Against the Valentinians* 7.5.

[91] Thomassen 2006:275–277 and 2000:5–6.

[92] Thomassen 2006:279–280 and 2000:13–14.

[93] Thomassen 2006:273–275, 283–287, and 2000:5, 9–12.

of the Dyad, just as it does in Moderatus' system. Thus the activities described in classical Valentinianism, rooted in the definition of number as progression, resonate with those of the neo-Pythagoreans. The monadic–dyadic tensions in Valentinianism echo the ambiguities and paradoxes seen in the contrast between the neo-Pythagorean systems discussed by Eudorus and Moderatus and the broader Platonist legacy. If there is any discernible departure from the neo-Pythagoreans, it is in the regular concern that plurality still needed to be explained and somehow linked to the prime mover. Valentinian theology criss-crosses to and from the one and the many.

Like Philo and Plutarch, the Valentinians used numbers to interpret texts and the world so as to illustrate religious truth. But whereas the first two were intent on illuminating tradition-based cults that were familiar to the public, the Valentinians were intent on explicating a private revelation within the Christian tradition—itself on the periphery of second-century Greco-Roman society. The Valentinians drew from competing metaphysical accounts offered by the neo-Pythagoreans and the Platonists, infused into these systems new arithmetical patterns and insights, and put their multiple levels of reality into story form. Christian concepts and entities played an important part, too, but the Valentinians chose to include them as parts of a larger mathematical universe. In one sense, Hippolytus was correct: intellectually, they were at heart Pythagoreans. But that Pythagoreanism was a literary tissue, not a community, an intellectual skin that the Valentinians stretched and altered as much as they depended upon it. With Marcus, as we shall see, that skin only grew.

4

The Apogee of
Valentinian Number Symbolism

Marcus "Magus"

THE MOST COMPLEX NUMBER SYMBOLISM of any Christian theology in the second century is found in the writings of a Valentinian named Marcus. He was given the epithet *magus* by ancient heresiologists because of his liturgical alchemy and his interest in ideas normally associated with magical texts. Very little is known about him. Förster, the only modern scholar to investigate Marcus's teaching thoroughly, suggests that he flourished between 160 and 180, which would make him a contemporary of other Valentinians such as Secundus and Heracleon.[1] He lived in Asia Minor, where he developed a cultic following within the churches. His teachings and liturgical practices agitated church leadership, which subsequently expelled him, a dispute immortalized in a polemical poem of the mid-second century, written by an unnamed orthodox cleric of Asia Minor (*Against Heresies* 1.15.6). In addition to eyewitness accounts and personal observation of a branch of Marcus' sect at work near his see of Lyons, Irenaeus had at his disposal texts that came from Marcus' circle, including a liturgy and an account of a revelation given to him. The revelation and Irenaeus' assorted paraphrases of Marcus' teaching reveal an esoteric, mystical arithmology that bordered on numerology.

Irenaeus begins his treatment of Marcus by revealing both the secret liturgical rites the latter used to seduce women into becoming his patrons and consorts, and the methods his followers used to draw away church members (*Against Heresies* 1.13). After describing the Marcosians' activities, Irenaeus introduces a revelation said to be from the Tetrad to Marcus (*Against Heresies* 1.14.1). That this is a paraphrase or close reconstruction of a written composition is indicated by Irenaeus' regular use of ancient quotation marks (the words "saying" and "says," λέγων, ἔφη) and, as we shall see, by the tight, self-referential

---

[1]   Förster 1999:390.

internal narrative. For the sake of convenience, I refer to Irenaeus' source as the *Revelation to Marcus*.

In the beginning of this text Marcus boasts that he has become the womb and receptacle of Colarbasus' Silence, that he is the Only Begotten and "most alone" (μονώτατος), and that he has brought forth the seed planted in him.[2] Thus Marcus identifies himself as the aeon traditionally placed third in the Valentinian Ogdoad: the Only Begotten, the offspring of Silence. He says that he is a kind of surrogate mother to the aeons.[3] As the *Revelation* progresses this relationship is enhanced, as the Tetrad descends to Marcus in the form not of a man but of a woman, since its masculine form would overwhelm the world. She tells him who she is, then reveals to him, whom she calls the "most alone," the creation of the universe, a revelation never before delivered to gods or people. Her cosmology is couched in obscure, difficult language, and is packed with detail. A cursory summary of the *Revelation to Marcus* (*Against Heresies* 1.14–16) would do no justice to the complex doctrines already compressed in Irenaeus' summary, and any strict translation is impenetrable. So to strike a balance between intelligibility and analysis, I paraphrase each passage (the original Greek appears in the Appendix, p. 191). I omit a number of details that do not materially affect my discussion of Marcus' number symbolism. Such omissions are few, since the symbolism is so pervasive. At the end of this chapter I synthesize the disjointed ends of the *Revelation to Marcus*, to assess Marcus' place in the early Christian theology of arithmetic.

Select Paraphrase and Analysis of the *Revelation to Marcus*

The Father—who is neither male nor female, and who is without substance and unknown—wished to make the unutterable utterable and to give shape to the unseen, and so opened his mouth and sent forth a Word similar to himself. The Word then came beside the Father and showed him who he was, becoming manifest as the shape of the unseen. The utterance of the name [*presumably the Father's name, which seems equivalent to the Word*] began with the Father speaking the first word [*or* Word], a collection (literally, "syllable") of four oral letters (στοιχεῖα): ἀρχή. He added a second collection, and it too consisted of four oral

---

[2] The epithet μονώτατος may allude to the terminology used at *Against Heresies* 1.15.1, where the highest aeon is called μονότης. See n. 25 below.

[3] It is impossible to tell whether the "of" in "womb of Silence" is objective or subjective (Förster 1999:166–167). He claims to be either a womb for Silence to be born on earth, or else Silence's personal womb, whereby her offspring can be brought to earth.

letters. Next, he uttered a third collection, of ten oral letters, and a fourth, of twelve. Thus there were four collections, a total of thirty oral letters. Each oral letter has its own written letters (γράμματα), impression, utterance, shapes, and images.[4] No oral letter sees or knows the form of the one of which it is an element. In their individuality, the oral letters know only their own utterance, and not their neighbors', and when they utter everything, the individual oral letters think they are naming the whole. These oral letters—parts of the whole—never stop echoing until there subsists only the last written letter of the last oral letter, speaking alone. That is the recapitulation, when everything, descending into one written letter, will resound with a single utterance. The image of this recapitulation is the word *amen*, when spoken by all of us in unison. The sounds provide shape to the uppermost aeon, which is without substance and is unbegotten.

(*Against Heresies* [*Revelation to Marcus*] 1.14.1)

My paraphrase of 1.14.1 is only slightly less confusing and crabbed than the original. The key idea here is that the Father utters a Word, which becomes the form of the invisible, much as happens to the Son in the *Tripartite Tractate*.[5] These two texts are similar in many other respects, except that the Father in the *Revelation to Marcus* first wills to be heard and seen and then opens his mouth, a process that emphasizes the monistic solitude of the Father and the adventitious character of the Word. We have no sense with Marcus that the Word is eternally preexistent in the Father. The Word is the Father's name, another striking parallel to the *Tripartite Tractate*.[6] This Word comprises a series of letter sounds in a pattern of four–four–ten–twelve, a pattern already familiar to us from the Valentinian Triacontad. But here the thirty aeons are separate from the Father and his Word. Whereas in other Valentinian systems the uppermost two entities are part of the Pleroma, here the Pleroma is constitutive of the Word, apart from the Father. Calling the entities 'letters' rather than 'aeons'

---

[4] I translate στοιχεῖον as "oral letter" and γράμμα as "written letter," since the *Revelation to Marcus* distinguishes the terms (Förster 1999:201), as did grammarians of the second century. See OCD s.v. "Dionysius (15) Thrax" with Uhlig and Hilgard 1883–1901:1:9; OCD s.v. "Apollonius (13) Dyscolus," with Uhlig and Hilgard 1883–1901:3:31–32, 323 (assigned to Apollonius Dyscolus at Schneider 1910:3). See also *Scholia in Dionysius Thrax* 1:323.33–35 (author: "Heliodoros"); 1.3:32.18–20, 1.3:31.19 (author: "Melampus/Diomedes"); 1.3:192.27–28 (author: "Stephen"). Each grammarian had his own scheme, but all distinguish a letter written from a letter uttered, even if the term στοιχεῖον was also used in a broader sense, to apply to letters' shapes or names (see e.g. *Scholia in Dionysius Thrax* 1.3:317.32–37). See also Förster 1999:198–199, 204, and p. 144 below.

[5] See e.g. NH 1.5:66.13–29.

[6] References and analysis at Thomassen 2006:180–181.

alters the imagery of the Pleroma: no longer are we dealing simply with powers and numbers. These are strings of utterances, blithely ignorant of each other, like the lines of detached, arcane letters and syllables one was to speak when casting magic spells, so common in late antiquity.

At the outset, the *Revelation to Marcus* creates a universe built upon linguistics, distinguishing uttered letters from their lower, written counterparts. Each oral letter has written letters and other properties attached to it. The oral letters are isolated from each other, and to reach their original unity they converge on a single oral letter that has a single written letter—an *aperçu* of the eventual resolution of all things in the Father. So the cosmology of the *Revelation to Marcus* begins and ends monadically, starting and terminating in a single utterance. And the world of the ear takes precedence over that of the eye.

> The regular, verbal names of the oral letters the Tetrad terms 'aeons,' 'words,' 'roots,' 'seed,' 'pleromas,' and 'fruit' [*that there are six terms anticipates other sixes to come*]. Every single name and its properties are encompassed by and understood through the name of the Church. Of these various oral letters, the last written letter of the last oral letter sends forth its own voice, the echo of which begets its own oral letters, which adorn the present world and the regions just above it. This last written letter is taken by its collection into the fulfillment of the All, but its echo, following along with the lower echo, remains exiled in the lower realms. The very oral letter whose written letter descended with her utterance contains thirty written letters, and each of these possesses other written letters, by which its name is declared, as well as the names of the letters composing its name, and so on. For example, the oral letter delta has five written letters: *delta, epsilon, lambda, tau,* and *alpha* ($\delta + \varepsilon + \lambda + \tau + \alpha$). And these five written letters are written down through other written letters, following the same process of begetting and succeeding, a process that can be extended infinitely. If this infinite expansion happens with a single written letter, how much greater the ocean of all the written letters belonging to that oral letter? How much greater for all the written letters of the entire name of Depth, the letters that the Forefather is composed of? So the Father, recognizing the Forefather's incomprehensibility, granted to each oral letter, unable on its own to utter the All, its own utterance.
>
> (*Against Heresies* [*Revelation to Marcus*] 1.14.2)

Details in this and the previous sections are mutually enlightening. Each of the thirty oral letters that form the Forefather's name contains written letters.

Note here how the *Revelation to Marcus* has moved from the names 'Father' and 'Word' to 'Depth'/'Forefather' and 'Father,' indicating a shift in protology as well. The thirtieth oral letter contains thirty written letters, and, so it is implied, the other twenty-nine oral letters contain thirty written letters each. Each written letter, when spelled out (e.g. δέλτα instead of δ), is composed of other written letters, which are themselves composed of others, and so on. The result is a three-tiered hierarchy. At the top is a triacontad of oral letters, each comprising a triacontad of written letters, the middle tier. Each of those written letters yields an ocean of written letters, thanks to the process of naming, the very process that caused Word to emerge from the Father. Thus the *Revelation to Marcus* posits an expansion of the Pleroma into a third, infinite realm, similar to the appearance of the innumerable aeons that populate the Church, the third tier in the Pleroma of the *Tripartite Tractate* (NH 1.5:70.24).

In other Valentinian systems Wisdom, the thirtieth aeon, breaks away from her consort, the twenty-ninth. In Marcus' revelation, the wisdom figure is the thirtieth written letter (or the last of the infinite chain of written letters generated by the thirtieth written letter), who breaks away from the thirtieth oral letter, to which she is attached. She gives voice, but produces merely an echo, and this echo produces a series of oral letters that resemble the highest sequence of oral letters, but lead to the creation of the demiurgic and material realms. The stray written letter is eventually restored, but the echo and its lower echo remain.[7] So the *Revelation to Marcus* consistently applies the grammatical analogy to the story of the aeonic fall and the creation of our lower world.

The twist on the Valentinian story produces other novelties. Sexual and gender metaphors have given way to grammatical ones. Aurality is to the visuality of a written letter as male is to female in the traditional Valentinian syzygy. Sound rules over image. This explains the sequence of 14.1, where the Father first wills the unutterable to be uttered, and only then the invisible to take shape. The hierarchy also explains why there is a limited number of oral letters (thirty), but an infinitude of written letters. The new metaphor sharpens in new ways certain points made in earlier Valentinian systems. Written letters are properties of spoken letters, and so cannot leave the root of their being. Hence the errant written letter's abortive attempt to give voice (φωνή) results in a mere echo and its own exile. The notion of a written figure trying to speak is as stark an absurdity as a woman trying to bear a child alone—both metaphors that were applied to the errant aeon Wisdom.

---

[7]   ἦχος here probably puns on Ἀχαμώθ, the entity called Resolution in Irenaeus' first Valentinian system.

> The Tetrad [*here called the* Tetraktys] now introduces Truth, depicted as a naked woman. Each of her twelve body parts is marked by two Greek letters: *alpha* and *omega* are assigned to her head, *beta* and *psi* to her neck, and so on down to her feet, *mu* and *nu*. This is the body of Truth, the shape of the oral letter and the impression of the written letter. The oral letter is called Human, who is the fount of every word, the source of every voice, the utterance of everything unspoken, and the mouth of mute Silence.

(*Against Heresies* [*Revelation to Marcus*] 1.14.3)

This section has little that pertains directly to number symbolism, but it is worth pointing out that the arrangement of the twenty-four Greek letters along two sides of a human figure was common in ancient astrology.[8] The *Revelation to Marcus* incorporates this image into its larger linguistic theory, portraying Truth's body as the extension of an oral letter and a written one. But this idea is not integrated into the larger cosmological model of the previous two sections, a sign of the *Revelation*'s eclectic nature.

> After the Tetrad's speech, Truth utters a word [or Word], which becomes a name, the name, the Tetrad says, "we know and speak: 'Jesus Christ.'" [*This is all Truth says throughout the Revelation.*] Seeing that Marcus expects her to speak further, the Tetrad intervenes and explains that this name, which she thinks Marcus might disparage, he does not adequately possess in its ancient form. She says that Marcus has only the sound and not the power, a power evident in that 'Jesus' is a noteworthy (ἐπίσημον) name, since it consists of six written letters, and a name known by the elect. The name, being compound, in the presence of the aeons has a different shape and form and is known by its kinsmen, whose Greatnesses are always with him.

(*Against Heresies* [*Revelation to Marcus*] 1.14.4)

Just as the Father uttered a word, which was his name, so Truth utters a word that becomes a name, implying that Jesus Christ is the express image of Truth.[9] The *Revelation to Marcus* focuses on the power of the name 'Jesus Christ,' and does so by calling on the obscure wordplay behind the term 'noteworthy' (ἐπίσημον). Both Marcus and Clement of Alexandria (as we shall see in chapter 7)

---

8   For the many primary sources and scholarly studies see Förster 1999:222–225 and Kalvesmaki, forthcoming.
9   Compare the *Gospel of Truth*, which calls Truth the Father's mouth and the Holy Spirit its tongue. NH 1.3:26.28–27.7.

made prominent use of the ἐπίσημον to make a recondite but lofty theological point, one that relies upon number symbolism for its impact.

The term ἐπίσημον must be addressed first, to correct confusion regarding the terminology of Greek numerals, and to set the stage for Marcus' arithmological *theologoumenon*. Greek numeration from the Hellenistic period onward followed an alphabetic convention, with α used for 1, β for 2, and so on up to ι, representing 10; κ for 20, and so on up to ρ for 100, σ for 200, and so on. To mark a letter as a numeral, a supralinear stroke was often placed just above it or to its right. Thousands were designated by the letters α through θ, placed usually at the beginning of the numeral and marked by a sublinear stroke to the left. Designating numbers in this fashion requires twenty-seven unique characters. But the classical Greek alphabet has only twenty-four. The three extra numerals were represented by nonalphabetic characters, whose placement in the series reflects their origin in a preclassical Greek alphabet from Asia Minor. The entire alphabetic system of numeration, with the nonalphabetic numerals underlined, is as follows:

α	1		ι	10		ρ	100
β	2		κ	20		σ	200
γ	3		λ	30		τ	300
δ	4		μ	40		υ	400
ε	5		ν	50		φ	500
ς	6		ξ	60		χ	600
ζ	7		ο	70		ψ	700
η	8		π	80		ω	800
θ	9		ϙ or ϟ	90		ϡ	900

Jesus' six-lettered name is called ἐπίσημον because this was the late antique term for ς, the Greek numeral six. That character is known most often today either as *stigma*, because of its resemblance to the ligature formed by *sigma* and *tau* (Ϲͳ), or as *digamma*, after the archaic Greek letter ϝ, which in some preclassical Greek alphabets was the sixth letter in the sequence, just as its archetype, the *waw*, was for the Phoenicians. But these are modern terms.[10] The preferred late

---

[10] *Digamma* is an ancient term, but as I argue below (p. 144), it referred to the obsolete letter, not the numeral. I have been unable to find the term *stigma* used in any ancient text. The same

antique and Byzantine term for the nonalphabetic numeral six was ἐπίσημον; as a class the nonalphabetic numerals were called παράσημα. An obscure but colorful late antique treatise, *The Mystery of the Letters*, calls Jesus the ἐπίσημον because the numeral six represents an empty space in the Greek alphabet, symbolic of the philosophers' rejection of Christ.[11] In that same text, Father and the Holy Spirit are symbolized by the κώφ (ϙ) and the ἐννακοσία (ϡ).[12] In an anonymous, undated treatise found in a late sixteenth-century manuscript, each of the three nonalphabetic numerals is named: ϡ is called χαρακτήρ; ϙ, σκόπητα; and ϛ, ἐπίσημος.[13] In various scholia on Dionysius Thrax written probably in late antiquity, the three signs are collectively called παράσημα, but are not individually named.[14]

The term ἐπίσημον implies something written or etched, but not uttered. Indeed, ἐπίσημον, παράσημον, and their cognates were widely used to describe imprints or other distinguishing marks on coins or shields.[15] Thus by calling Jesus the ἐπίσημον, the *Revelation to Marcus* uses number symbolism to highlight, through the six letters of his name, both his "excellence" and the way he imprinted himself upon the coin of humanity. It also emphasizes the humbleness of the Incarnation, since Jesus deigned to become written, not uttered, letters. I say much more about the ἐπίσημον in chapter 7.

> The twenty-four written letters of the Greek alphabet are "reflective effluences" of the three powers that encompass the entire number of the upper oral letters. There are nine consonants (ἄφωνα), which correspond to the Father and Truth because they too lack sound [ἀφώνους, *a*

---

    applies to the term *sampi* for ϡ. Surely, this derives from the Byzantine expression [ὡ]ς ἂν πῖ ("just like pi"), but the only attempt I have found to date the term is that of Keil: "Dieser Name [Sampi] stammt übrigens in dieser Form aus der 2. Hälfte des 17. Jahrh. n. Chr." (1894:265n2). But Keil gives no evidence.

[11] *Mystery of the Letters* 2:30–31, 33 (Bandt 2007:108, 170–176). For more on this unusual text, see Dupont-Sommer 1946; Galtier 1902; Bitton-Ashkelony 2007; and Kalvesmaki, forthcoming.

[12] *Mystery of the Letters* 33 (Bandt 2007:174). A similar list appears in a ninth-century codex, the *Psalterium Cusanum*, in a Latin text intended to acquaint readers with Greek conventions, but with the last two terms mixed up: S (VI) is *Episimōn*, F (XC) is *Enacōse*, and ϡ (DCCCC) *Cophē*: cod. 9, fol. 64v in Marx 1905:6–7. The manuscript is discussed in Gardthausen 1913:260 and C. Hamann 1891.

[13] Vienna, MS theol. gr. 289, fol. 44r. See Hunger and Lackner 1992:s.v.

[14] This appears in two very similar passages, attributed to different authors (a certain Heliodorus, and anonymous): Uhlig 1883:1.318.29–37 and 319.21–31. The idea of the three numerals as "signs" is continued in the Latin and Greek manuscript *Laon*, cod. 444, f. 311v, column a: "⊦ et ⊣ et ϛ et F et ↑ *non sunt literae apud Graecos, sed notae et signa*" (*Catalogue général* 1849:234–236). See Miller 1880:213. The term ἐπίσημον fell out of currency not long ago. In the early eighteenth century Montfaucon called the three nonalphabetic numerals ἐπίσημον βαῦ, ἐπίσημον κοφῆ, and ἐπίσημον σαμπι (1708:122, 128, 132).

[15] LSJ, 655b–656a, 1323b–1324a. Note, however, that LSJ does not include the technical definition of παράσημα discussed here.

pun that turns on φωνή 'voice']. The eight semivowels (ἡμίφωνα) reflect Word and Life, since they dwell between the vowels and consonants and can absorb the effluence of the consonants coming from above and the reciprocation of the vowels coming from below. The seven vowels belong to Human and Church, since the echo "of his voice" gave shape to the All.[16] Because of the unequal ratio 7:8:9, one of the nine who was enthroned beside the Father descended on a mission to the realm that was once off limits to him, to rectify what had been done, so that the unity of the Pleroma, which possesses equality, might generate from the All a single power for the All. From this embassy, the realm that has the seven vowels obtains the power of eight, thus equalizing all three realms at eight members apiece. All three are then ogdoads, and in their interplay the three groups of eight furnish evidence for the number twenty-four. [*At this point the* Revelation to Marcus *provides another story, about the generation of the lower oral letters. Unfortunately, the two sentences explaining this are garbled. The various translations are generally accurate, but unintelligible.[17] What is clear is the following:*] There are three oral letters who belong to the three Powers [*mentioned above as encompassing the All*]. Because the three Powers are syzygies, the three oral letters are actually six.[18] The twenty-four oral letters flow out of the three oral double letters. When the three oral double letters are quadrupled by the ratio, factor, or word of the ineffable Tetrad, they create a number equal to the aforementioned twenty-four oral letters. These twenty-four oral letters belong to the Unnamed One. The three oral letters are worn or carried by the three powers, to imitate the invisible one. The double written letters in the Greek alphabet (ζ, ξ, ψ) are an image of these three oral letters, which are themselves images [i.e. of the three Powers]. When the three double oral letters are added to the twenty-four oral letters, it makes the number thirty, through its potential for proportion.[19] [*No wonder the translations are unintelligible.*]

(*Against Heresies* [*Revelation to Marcus*] 1.14.5)

---

[16] The nine consonants: π, κ, τ, β, γ, δ, φ, χ, θ; eight semivowels: λ, μ, ν, ρ, ς, ζ, ξ, ψ; seven vowels: α, ε, η, ι, ο, υ, ω. This threefold division of the alphabet is typical in this period. See Förster 1999:238–242 for parallels and discussion.

[17] E.g. Rousseau et al. 1965–1982:2.223, Förster 1999:234; ANF 1.337, 5.95; F. Williams 1987:217.

[18] Förster sees this point as unclear, and offers different solutions (1999:247–248). His alternative suggestion (248, paragraph 2), that this refers to the double letters ζ, ξ, ψ, is most plausible. We need not infer that these were to be written out ΔΣ, ΚΣ, ΠΣ, as Förster suggests, since these are στοιχεῖα, not γράμματα (see n4 above).

[19] The phrase δυνάμει τῇ κατὰ ἀναλογίαν is peculiar. The closest parallel I have found is in Alexander *Commentary on Aristotle's "Metaphysics,"* 682.19–20 (ed. Hayduck 1891), where commensurability is qualified as being either potential or actual.

The key to understanding this section is to appreciate its setting. We are no longer in the Pleroma, but outside it, beginning with the realm of three Powers who circumscribe the Pleroma, and ending with what seems to be the origin of the Greek alphabet. The entire section explains how the alphabet ultimately derives from the Powers, but the two halves of the text reverse the proper narrative order. The three Powers who circumscribe the All are syzygies. How these Powers fit into the larger scheme is never explained, although it is tempting to see them as equivalent to the six-sided Limit that is found in Irenaeus' first Valentinian system, which also surrounds the Pleroma. Or the three Powers may simply be the Hexad in Hippolytus' Valentinian system, or the lower three of the four syzygies that make up the Ogdoad. No matter. The three Powers utter oral letters, six in total, which are their image. The six oral letters are multiplied by the ineffable Tetrad [another figure whose position in the system is not explained] to project twenty-four oral letters that belong to the Unnamed One (again, an unexplained entity; it cannot be any of the characters we have met so far, since they all, including the ineffable Father, have names). All these oral letters combined are potentially proportionate to the upper thirty oral letters. Now, just as those upper oral letters have derivative written letters attached to them, so too have the lower oral letters. The lower, derivative written letters are the Greek alphabet, and the three classes of letters, like the three Powers, are images of the three bottom pairs of the primal Ogdoad, correlated in a hierarchy that begins with consonants, opposite to what a Platonist might have expected:[20]

Father and Truth	nine consonants
Word and Life	eight semivowels
Human and Church	seven vowels

The thirty lower written letters are said to be potentially proportionate to the upper ones, and here, in the derivative twenty-four written letters, we see why. The proportion of consonants to semivowels to vowels is out of step with the equality exhibited in the Ogdoad. To rectify this inequality, there descends yet another entity whose origin and status are murky, and who is described in contradictory language.[21] He effects a transfer of one consonant to the vowels,

---

[20] See *Philebus* 18bc on the Egyptian origin of the three classes, with vowels at the top of the hierarchy.

[21] The key phrase is ὁ ἀφεδρασθεὶς ἐν τῷ πατρὶ κατῆλθεν. The hapax legomenon ἀφεδράζειν may be as innocuous as Lampe takes it—'to set apart' (1961: s.v.)—but the Greek words most closely

so that all three classes equally become Ogdoads, and thereby point to their corresponding twenty-four oral letters, which were generated evenly by the product of the Tetrad and the three pairs of oral double letters. Which consonant made the transition? We are left to guess. Further, three of the lower written letters, namely the three Greek double consonants (ζ, ξ, ψ), are mere reflections of the three pairs of oral letters that emerge from the three Powers. That is, the six oral letters are not themselves necessarily the sounds /d//s/, /k//s/, and /p//s/; it is merely that the three Greek double consonants point to their arrangement. We do not know what actual sounds they—or for that matter any of the upper oral sounds—represent.

This section is confusing partly because there are several numerical series at work, series that would be mutually incompatible were they not unfolding on different levels of the cosmology. At the highest level is the series 4–4–10–12—oral letters constituting the Word (*Revelation to Marcus* 1.14.1). Each letter presides over thirty written letters, presumably grouped in similar fashion. Outside this highest realm, the All, are three syzygal Powers who generate a series of oral letters: 2–2–2, multiplied by the Tetrad to result in 8–8–8. Beneath this are the written Greek letters, which follow a series that is transformed from 9–8–7 to 8–8–8.

> Concerning this ratio [*that of the six oral letters quadrupled by the Tetrad*] and its providential arrangement, the Fruit [*another name for Jesus in Irenaeus' extended Valentinian system*] has appeared in the likeness of its image [cf. Romans 1.23]. This is he who, after six days, ascended the mountain as the fourth and became sixth [Matthew 17.1, Mark 9.2], who then descended and was thereby possessed by the Hebdomad. He was the "episēmos ogdoad"—the "noteworthy octet." He also possessed in himself the entire number of oral letters. When he came to be baptized, the descent of the dove made this number manifest [Matthew 3.13–17, Mark 1.9–11, Luke 3.21–22]. The sum of the letters in περιστερά 'dove' is 801, written as ωα'—the *omega* and the *alpha*. Because of all this, Moses said that man was created on the sixth day and the divine dispensation occurred in the sixth day [Genesis 1.31]. This is the Day of Preparation, when the last man was made manifest for the rebirth of the first. The beginning and end of this divine dispensation occurred at

---

related suggest it means to discard menstrual or toilet waste! See LSJ, s.v. ἄφεδρος, ἀφεδρών, ἀφεδρεία, and especially Matthew 15.17. Neither possibility can easily handle "in the Father," since the verbal prefix ἀπό contradicts ἐν. The Latin translator, otherwise quite literal, gave up here and rendered the entire phrase *qui erat apud Patrem descendit.* I believe the best interpretation is that this entity found itself in the lower regions, but was rescued, enthroned within the Father, and sequestered from the lower regions until called to this mission.

the sixth hour, when he was nailed to the wood [Matthew 27.45, Mark 15.33, Luke 23.44]. Why? Because the perfect Mind, knowing that the numeral for six possesses the power of creation and rebirth, revealed to the sons of light the rebirth that came about through the appearance in him of the number of the ἐπίσημον. This is why the double letters (ζ, ξ, ψ) refer to the ἐπίσημον.[22] For when this ἐπίσημον was mixed with the twenty-four oral letters, it perfected the name written with thirty letters.

(*Against Heresies* [*Revelation to Marcus*] 1.14.6)

The *Revelation to Marcus* interprets the Transfiguration as a numerical code for the construction of the universe. It starts with the six days, mentioned in two of the Gospel accounts of the Transfiguration, and takes it as a sign of the three doubled oral letters. Jesus, the fourth person (Peter, James, and John are the other three), represents the ineffable Tetrad. At the appearance of Moses and Elijah he becomes the sixth, the factor against which the Tetrad generates the twenty-four oral letters. So Jesus possesses within himself the entire number of the oral letters. His ascent, then, symbolizes the generation of the thirty oral letters from the three Powers, discussed in the previous section. Likewise, Jesus' descent from the mountain represents the movement from the oral to the written letters. He becomes a metaphor for the written letter that joins the hebdomad of vowels to convert it into an ogdoad.

The *Revelation to Marcus* draws upon the earlier epithet for Jesus, the ἐπίσημον, the numeral six, and combines it with this mission to create a new epithet, 'episēmos Ogdoad.' The paradox of the "sixly octet" has been alluded to earlier, in 1.14.4, where Truth utters only two words to Marcus: Χρειστὸν Ἰησοῦν, a name of eight and six letters.[23] The Tetraktys comes alongside Marcus and explains to him that one must go beyond the mere sound of the name and penetrate its power. The power of the name is this: 'Jesus' is the "noteworthy name" (ἐπίσημον ὄνομα) because it has six letters. 'Christ,' which consists of eight letters, has the power of the Ogdoad. Hence the epithet 'episēmos Ogdoad'.

---

22 That is, in addition to the cosmological scheme, discussed below, the double letters signify the number six: ζ is the sixth letter (despite its numerical value of 7), ξ = 60, and ψ = 600.

23 Although the editions of Hippolytus (*Refutation of All Heresies* 6.45.1.3) and Epiphanius (*Panarion* 2.12.20), upon which Irenaeus' Greek text is largely reconstructed (*Against Heresies* 1.14.4.6), render the name Χριστός (the spelling preferred today), the original has to have been Χρειστόν since, later in the text, Marcus makes the theological point that "Son, Christ" (Υἱός Χρειστός) is composed of twelve letters, and "Christ" (Χρειστός), of eight (Irenaeus *Against Heresies* 1.15.1.39, 2.41–50; cited in Hippolytus *Refutation of All Heresies* 6.49.4.4, 5.1; Epiphanius *Panarion* 2.18.11, 19.16–20.1). What is otherwise a rather innocuous variant in spelling (to the modern editor, a misspelling) here takes on theological importance. See n. 26 below.

Marcus also interprets the Baptism so as to link Jesus to the entire alphabet. The value of the letters in περιστερά 'dove' (80 + 5 + 100 + 10 + 200 + 300 + 5 + 100 + 1) is 801. This number is written ωα′, and therefore points to Christ as the *alpha* and the *omega*, the beginning and end of the Greek alphabet. Then follow four exegetical points concerning numbers. The first pair deals with man's creation: the first man was created on the sixth day of Creation, and was reborn by the last man in his passion, also on the sixth day. The second pair concerns the divine dispensation (οἰκονομία), which happened on the sixth day of the Transfiguration, but as a sort of transit point to and from the cross: the beginning and end of the dispensation is in the sixth hour, when Jesus was crucified. So Scripture places the power of the number six at the center of the Creation and its dispensation.

> Silence goes on and says that the ἐπίσημον uses the magnitude of seven as a deacon, so that Fruit, of its own independent will, might be made manifest. She charges Marcus to think of the numeral ἐπίσημον in current use as the one who was shaped into the ἐπίσημον, the one who was, as it were, divided in half and remained outside. This is the one who, through his power and forethought, endowed with a soul this world—the world of the seven powers, seven to imitate the power of the Hebdomad—and everything visible, by means of that which he projected. The ἐπίσημον uses the resultant product as if it had emerged on its own. Other matters, being imitations of inimitable things, serve the Ἐνθύμησις of the Mother. Each of the seven heavens that constitute this world utters a vowel, from *alpha* to *omega*, and the intermingled sound of the seven Powers echoes and glorifies the one who projected them, and the glory of the echo is offered up to the Forefather. The echo of that doxology descends to earth and thereupon molds and creates the things of earth.

> (*Against Heresies* [*Revelation to Marcus*] 1.14.7)

So much in this section is vague, ambiguous, or unexplained that it is difficult to flesh out the cosmogony accurately. It appears to describe how the entity Akhamoth created the world by means of her projection, who in turn created the cosmos and the seven heavens as a reflection of the Hebdomad, to give glory to the Forefather. The echoes of this glory become responsible for shaping the earth. The *Revelation to Marcus* alludes here to the Valentinian myth of the exiled aeon Akhamoth, representing him (not her!) by the number six, and her projection, the demiurge, by the number seven, the number used for the creation of this world. But at this point in the *Revelation to Marcus*, so many things have been called the ἐπίσημον—Jesus, the six oral letters projected by the three Powers, the Greek numeral, and now Akhamoth—and so many pronouns have been sundered from their referents that the overall picture is rather jumbled.

> The proof of this is the soul of a newborn baby, who, while proceeding from the womb, cries out the echo of each of these seven oral letters [*that is, the vowels*]. Just as the seven Powers glorify the Word, so the soul in wailing infants glorifies Marcus himself [*probably, but not certainly, Irenaeus' sarcastic interjection*]. All this is illustrated in David's phrases, "Out of the mouth of infants and nursing babes thou hast perfected praise" and "The heavens declare the glory of God" [*Psalms* 8.3, 18.1 LXX]. So too, the distressed soul often resorts to uttering the last vowel, *omega*, in times of distress, as a sign of praise, so that the higher soul, its kinsman, might dispatch help.
>
> (*Against Heresies* [*Revelation to Marcus*] 1.14.8)

Here we see the *Revelation to Marcus* connecting its cosmogony to the experience of human utterances, drawing from themes touched on at the end of 1.14.7. The seven vowels were a frequent point of interest both in general number symbolism and in magical spells, where the vowels would be written in pyramid fashion:[24]

$$\alpha$$
$$\varepsilon \ \varepsilon$$
$$\eta \ \eta \ \eta$$
$$\iota \ \iota \ \iota \ \iota$$
$$o \ o \ o \ o \ o$$
$$\upsilon \ \upsilon \ \upsilon \ \upsilon \ \upsilon \ \upsilon$$
$$\omega \ \omega \ \omega \ \omega \ \omega \ \omega \ \omega$$

The seven vowels were held to have an innate connection with the seven planets and the seven notes of the musical scale. This association, regarded as Pythagorean, no matter its historical merits, the *Revelation to Marcus* capitalizes on and justifies with Scripture, to lend support to its cosmogony. So Marcus' theological edifice straddles Scripture, Pythagorean lore, and occult science.

> [(1.14.9) *This section merely recapitulates 1.14.1-8.*] Concerning the Tetrad: Henotes coexists with Monotes, and from them come two projections, Monad and Hen. Two plus two make four, and when the operation is repeated—four is added to two—the number six is made evident, and these six quadrupled bring forth the twenty-four forms. Silence then turns to the names of the first Tetrad, to show how they couch within

---

[24] *Papyri Graecae Magicae* 1.13–19. Other examples are legion.

themselves ineffable, holy mysteries known only to the Father and Son. The names Ἄρρητος and Σειγή [*sic*], Πατήρ and Ἀλήθεια consist of a total of twenty-four oral letters, since the first and last each have seven written letters [*the distinction between στοιχεῖον and γράμμα now blurs*] and the middle two, five. The same can be shown in the second Tetrad, Λόγος and Ζωή, Ἄνθρωπος and Ἐκκλησία, the written letters of whose names add up to the same number. Further, the uttered name of the Savior, Ἰησοῦς, consists of six written letters, but his unutterable name has twenty-four written letters. Yet Υἱὸς Χρειστός [*sic*: "Son Christ"] has twelve written letters, so that the unutterable name in Christ has thirty written letters. This, says Silence, explains why he is *alpha* and *omega*, in order to disclose the dove, the letters in whose name add up to this number [*explained above*].

(Against Heresies [Revelation to Marcus] 1.15.1)

[In one manuscript, toward the end of this section, is an addition that probably comes from the *Revelation to Marcus* or a related early source. It goes as follows:]

καὶ αὐτὸ τοῖς ἐν αὐτῷ γράμμασι κατὰ ἓν στοιχεῖον ἀριθμούμενον, τὸ γὰρ χριστόν ἐστι στοιχείων ὀκτώ· τὸ μὲν γὰρ χρῑ τριῶν, τὸ δὲ ρ̄ δύο, καὶ τὸ εῖ δύο, καὶ ῑ τεσσάρων, τὸ σ̄ πέντε, καὶ τὸ τ̄ τριῶν, τὸ δὲ οῦ δύο, καὶ τὸ ν̄ τριῶν· οὕτως τὸ ἐν τῷ χριστῷ ἄρρητον φάσκουσι στοιχείων τρίακοντα.

[My paraphrase:] When the written letters in each oral letter in the name Χρεῖστος are added up you get thirty: *chi* has three letters, *rho* two, *ee* two, *iota* four, *sigma* five, *tau* three, *ou* two, and *nou* three. That is, χεῖ + ρό + εῖ + ἰότα + σῖγμα + ταῦ + οῦ + νοῦ [*my reconstruction*] is twenty-four letters, which when added to the unutterable name of six letters makes thirty.

Just as there was a slight shift in Valentinian protologies from 1.14.1 to 1.14.2, now there is another, more significant one, as the *Revelation to Marcus* begins anew with the system of Epiphanes, the distinguished Valentinian teacher discussed above in chapter 3.[25] Rather than explore the inner workings

---

[25] *Revelation to Marcus* 1.11.3, which Förster suggests is a literary fragment of Marcus (1999:15, 296). Besides the obvious parallels in substance, he says, καθὰ προείρηται in Irenaeus *Against Heresies* 1.16.1 seems to refer to 1.11.3. I accept the parallelism, but I doubt that the two passages are by the same author. Yes, both use the same names for the elements of the Tetrad: μονότης, ἑνότης, μονάς, ἕν. But at 1.11.3, the male members of the tetrad are also called ἀρχαί (or προαρχαί), whereas female members are called δυνάμεις. Marcus uses 'source' and 'power' without the terminological rigor of Epiphanes. Further, the source for 1.11.3 uses προίημι for "project," suggesting it was his preferred term (cf. Irenaeus' corresponding mockery, which thrice reuses the term, 1.11.4, lines 74 [bis] and 79 in Rousseau et al. 1965–1982). Marcus also uses the same

of Epiphanes' Tetrad, the text incorporates extra arithmetical procedures, to forge an affinity between this Tetrad and his theory of the emergence of the twenty-four oral letters. This provides a bridge to the classical Valentinian Ogdoad, by means of letter counts. No attempt is made to reconcile the metaphysical incompatabilities between Epiphanes and the standard Ogdoad.

The concept of naming, which was so dominant a theme in the early parts of the *Revelation to Marcus*, now returns, to explain the logic behind the names of the traditional Valentinian Ogdoad and Jesus Christ. The names are not arbitrary, as shown by the number of letters in each, provided one uses the correct spelling variation. Thus, the *Revelation to Marcus* depends upon the longer spelling of 'Christ,' and, within that spelling, upon a creative spelling of its constituent alphabetic letters.[26]

> Jesus' ineffable origin is shown by the generation of the second Tetrad, which comes forth from the first as if a daughter from a mother. They become an ogdoad and from this emerges the Decad. The Decad comes alongside the ogdoad, multiplies it by 10, and makes it 80. Multiplying by 10 once more, the 80 becomes 800. Thus the entire number of written letters is demonstrated by the progression from ogdoad to decad, from 8 to 80 to 800, for a total of 888, the value of the sum of the letters in Ἰησοῦς. This clearly indicates that Jesus' birth is supercelestial. It also explains why the Greek alphabet has eight units, eight tens, and eight hundreds: it points to Jesus, who consists of this number. And it explains yet again why Jesus is named the *alpha* and *omega*, to indicate his origin in all things. Furthermore, when the first Tetrad was added incrementally to itself [1 + 2 + 3 + 4], the number 10 appeared, and this is represented by *iota*, Jesus' initial. Further, Χρειστός has eight written letters and thus indicates the first Ogdoad, which, in combination with 10, produces 'Jesus' [i.e. 8 + 10 = η´ + ι´ = ιη´(σους)]. He is also called Υἱὸς Χρειστός, which has four plus eight written letters, thus indicating the magnitude of the Dodecad. Before the ἐπίσημον of his name, 'Jesus,' appeared, people were in exceeding ignorance and error. But at the appearance and incarnation of the six-letter name, which possesses both the six and the twenty-four, those who knew this were freed from ignorance and went from death

---

verb, but without the same regularity, and only twice (1.14.1 [line 143 Greek], 1.14.2 [line 176 Greek]). Thus I treat the two passages as independent, with Marcus adding Epiphanes to his bricolage.

[26] Other texts rely upon unconventional spellings of Χριστός to make specific points. See n. 23 above, and Förster 1999:318–319; Hippolytus *Refutation of All Heresies* 6.49.4–5; and an inscription at Shnān (IGLS 1403, with commentary by Kalvesmaki 2007a).

to life. His name becomes for them a path to the Father of Truth, who desired to destroy ignorance and death, so he could be known. This is why the human constructed in the image of the upper Power was elected.

(Against Heresies [Revelation to Marcus] 1.15.2)

So the mathematical properties of the alphabet and of the Pleroma are demonstrations of the power in the name 'Jesus.' The sum of the numerical values of the letters making up the name Ἰησοῦς is 888 (10 + 8 + 200 + 70 + 400 + 200), a number whose significance is confirmed by the alphabetic system of numeration and the interplay of the Ogdoad and Decad. That interplay is further paralleled in the coincidence of the initial *iota* and the eight letters of 'Christ,' the two parts of the familiar name 'Jesus Christ.' The less common title 'Son Christ' points to the Dodecad. All in all, the *Revelation to Marcus* argues that the six-letter name 'Jesus'—stressing the name over the person—is a conduit to knowledge of the Father. Note that this section conceives of the Ogdoad as consisting of one tetrad begetting a second, but it builds upon that familiar model to arrive not at the Triacontad but at explanations for the name Jesus and for the structures of the Greek alphabetic numerals.[27]

> The aeons, or powers, proceed from the Tetrad formed by Human and Church, and Word and Life. These powers generate Jesus, whose appearance on earth consists of four places, reserved for Word, Life, Human, and Church, supplied by the angel Gabriel, the Holy Spirit, the power of the Most High, and the Virgin respectively (Luke 1.26, 35). The human born by divine dispensation through Mary was chosen by the Father, to make himself known by means of Word, and when Jesus entered the water there descended upon him as a dove the very power who ascended and fulfilled the twelfth number. This power is the Father's seed, and in him, sown at the same time, is the seed of all of who descend and ascend with him. This power possesses within himself Father, Son, the unnamed power of Silence, and all the aeons. This power is the Spirit who speaks in Jesus, confesses him to be the Son of Human, and makes the Father manifest. It descended on Jesus and united with him. This, the Savior of the divine dispensation, destroyed death and reveals that Christ is his father. Although the name of the (ordinary) human chosen for the divine dispensation was 'Jesus,' the name was fashioned in the resemblance and shape of the Human who

---

[27] Compare the several Tetrad-to-Tetrad models of the Ogdoad reported by Irenaeus at *Against Heresies* 1.11.

was about to descend on him. This Human contained the Ogdoad: Human, Word, Father, Ineffable, Silence, Truth, Church, and Life.

(*Against Heresies* [*Revelation to Marcus*] 1.15.3)

This section describes the mission of the Savior and distinguishes sharply between the human born of Mary and the Human sent by the Ogdoad of the Pleroma to unite with him at baptism. To prepare for this union the second Tetrad issued four powers to create four spaces within this human, adequate for the salvific mission. Just as in Irenaeus' extended Valentinian system, this savior figure is a Tetradic being, in imitation of the Pleroma. In Marcus' system, however, all four elements of Jesus are assembled by the second Tetrad, not the first. His Tetradic constitution also explains his ascent to the mountain as the fourth person.

In this section there is a slightly obscure response to the criticism that emphasizing the numerical properties of the letters of a pedestrian name like 'Jesus' was ridiculous. The *Revelation to Marcus* grants that the name was mundane, but says that it nevertheless took on special supernatural properties, actualized by his union with the Savior figure. The argument of this section, in line with previous parts of the revelation, is that when one looks beneath the surface at the mathematical symbolism, the name 'Jesus' is very powerful, as revealed in its number and letter symbolism.

[(1.15.4–6) *In these sections Irenaeus either criticizes Marcus' doctrines or merely recapitulates earlier material. When Irenaeus returns to Marcus' teaching, it is no longer obvious that he is depending upon the Revelation to Marcus as his source. When he mentions his opponents he moves from the singular to the plural, which suggests that the remainder of his report is based upon material derived from a number of Valentinians, Marcus included.*] Those people, who reduce everything to numbers, try to combine the unfolding of the aeons mystically with the parable of the lost sheep. They claim that all things come from Monad and Dyad. Adding up from Monad to four [1 + 2 + 3 + 4], they engender the Decad. The Dyad, too, progressing from itself up to the ἐπίσημον [2 + 4 + 6] indicates the Dodecad. Yet a further progression of even numbers from the Dyad to the number ten reveals the Triacontad [2 + 4 + 6 + 8 + 10], wherein reside the Ogdoad, Decad, and Dodecad. This Dodecad they also name Passion, since it has the ἐπίσημον accompanying it.[28] This is expressly related to the fall of the twelfth member of the Dodecad, and this event is related to the parables of

---

[28] The various Greek editions of Irenaeus, Hippolytus, and Epiphanius (in Rousseau et al. 1979, Marcovich 1986, and Holl 1922, respectively) make little sense. The three Greek manuscripts

the lost sheep and the lost drachma [Luke 15.1–10]. In the former parable the grazing sheep wandered off; in the latter, the woman who lost her drachma lit a lamp and found it. In the former parable there were eleven members left over, and in the latter, nine. When these remainders are multiplied [9 × 11], they give birth to the number ninety-nine. For this very reason Amen has this same number [Ἀμήν = 1 + 40 + 8 + 50].

(*Against Heresies* [*Revelation to Marcus*] 1.16.1)

The Valentinians appeal in this section to the mathematical properties of the Tetraktys (see Excursus B) and the series of even numbers to explain the logic intrinsic to the thirty-aeon system. To the same end, to show how the Triacontad is encoded in Scripture, they then point to two parables that mention numbers. In one of these parables, concerning the lost sheep, the substitution of ninety-nine for twelve may seem baffling, but it is explained in the next section.

The reference to *amen* and its psephic sum is significant. This is the earliest association of *amen* with the Greek numeral for 99, ϙθ′, which appears in numerous manuscripts and papyri from late antiquity and the Byzantine period as an abbreviation for *amen*.[29]

The oral letter *eta*, along with the ἐπίσημον, is an ogdoad, since it is in the eighth place after alpha. Furthermore, reckoning the oral letters up to *eta*, without the ἐπίσημον [1 + 2 + 3 + 4 + 5 + 7 + 8], adds up to the number thirty, thereby revealing the Triacontad. This shows that the Ogdoad is the "mother of the thirty aeons." And since the number thirty is compiled from three powers [i.e. Marcus' three Powers], so when it is tripled it makes ninety. So the Trinity, multiplied by itself,

---

(P = Paris. supp. gr. 464 [14th c.], V = Vat. gr. 503 [9th c.], M = Marcianus 125 [11th c.]—commas omitted), compared with the Latin text edited in Rousseau et al. 1979:

> *Duodecadem igitur eo quod episemon habuerit consequentem sibi propter episemum, passionem vocant.*

P   τὴν οὖν δωδεκάδα διὰ τὸ ἐπίσημ(ον) ἐσχηκέναι συνεπηκολούθησεν αὐτῇ τὸ ἐπίσημον πάθος.

V   τὴν οὖν δωδεκάδα διὰ τὸν ἐπίσημον διὰ τὸ συνεσχηκέναι συνεπακολουθήσασαν αὐτῇ τὸ ἐπίσημον πάθος λέγουσι.

M   τὴν οὖν δωδεκάδα διὰ τὸν ἐπίσημον διὰ τὸ συνεσχηκέναι συνεπακολουθῆσαν αὐτῇ τὸ ἐπίσημον πάθος λέγουσι.

Based on its affinity with the Latin, M seems the superior reading, although we may wish to emend the ninth word to read {συν}εσχηκέναι. M follows the Latin almost precisely, with the notable exception that *propter episemum* is placed after the *eo-quod* clause, not before, where it could be taken as the referent of a relative clause. The thrust of the passage is that the Dodecad is being given the epithet 'Passion,' and this because of the action of the ἐπίσημον.

[29]   Among many studies see Robert 1960, Vidman 1975. The intriguing question of whether Marcus' *theologoumenon* was directly responsible for the widespread phenomenon remains open.

makes nine. In this way the Ogdoad gives birth to the number ninety-nine. So they say that the twelfth aeon abandoning the Dodecad was a type of the written letters that are positioned in the arrangement of the Word. That is, the eleventh written letter is Λ, *lambda*, whose numerical value, thirty, remains an image of the upper, divine dispensation, since the sum of the written letters that precede it [*again, omitting the* ἐπίσημον] is ninety-nine. That lambda, being eleventh in rank, descended to find the one similar to it, to complete the twelfth number. Once the lambda found it, it was completed by the shape of the oral letter. The letter M [the twelfth letter] is made up of two Λs lying side by side. Therefore they use knowledge to flee the land of the ninety-nine to pursue the one (τὸ ἕν), which, when added to the ninety-nine, results in a transfer from the left hand to the right.

(*Against Heresies* [*Revelation to Marcus*] 1.16.2)

In this final section Irenaeus reports yet more demonstrations of Valentinian logic based on the alphabet, alphabetic numeration, and the shapes of the letters. The Ogdoad leads to the Triacontad. The Triacontad is linked to the three Powers, called here a Trinity (τριάς, the same term used by the orthodox for Father, Son, and Holy Spirit), which leads to the ninety-nine. This calls to mind the ninety-nine sheep in the parable discussed in the previous section and in the *Gospel of Truth* (see chapter 3 above). Here is the explanation for why the Valentinians felt free to use eleven instead of ninety-nine as the number symbolizing the sheep who were not lost: $α + β + γ + δ + ε + ζ + η + θ + ι + κ + λ$ (the first eleven letters) equals $1 + 2 + 3 + 4 + 5 + 7 + 8 + 9 + 10 + 20 + 30$, which equals 99. Note also that *lambda* is the first letter of Λόγος, the "Word" referred to in the previous sentence. All this, plus one *lambda*'s rescuing the other, make for a colorful addition to the Valentinian interpretation of the parable of the lost sheep via finger counting.

As should be evident from my paraphrase and commentary, Marcus' theology is marked by a surfeit of number symbolism, now combined with the symbolism of letters and grammar. No other Valentinian system shows such an extensive, elaborate use of numbers. Marcus draws eclectically from various Valentinian protologies, not bothering to reconcile contradictions. He seems not to have cared about the lack of philosophical purity of some of his predecessors. This is illustrated well by asking whether he was a monadic or dyadic Valentinian. On one hand it would seem that he is extremely monadic. He specifies that the Father has no gender and exists alone, even before the emanation of his Word (*Against Heresies* 1.14.1). The identification of the upper Tetrad as consisting of four different kinds of unity, in line with Epiphanes, reinforces this monadic ideal (*Against Heresies* 1.15.1).[30] The human search for unity, epitomized in the mathematical and linguistic return to a single letter and sound, reinforces this monism (*Against Heresies* 1.14.1). Elsewhere (1.13), Irenaeus accuses Marcus of telling his women adherents "we must become as one" (τὸ ἕν), a formula repeated three times. The prayer uttered by his followers presumes that they have achieved unity with a certain intermediary being, a counselor of God and Silence. Collectively, these references suggest that Marcus envisioned a metaphysical unity or monism as the beginning and the ultimate goal of life.

But the same prayer that yearns for monadic unity also assumes that the primal being and his pre-eternal consort, Silence, are naturally paired. Marcus stresses the importance of the uppermost conjugal bond when he claims that Truth, the fourth aeon, the "source of every word and every voice"—probably an allusion to the fifth and sixth aeons—is the projection of Ineffable and Silence (*Against Heresies* 1.14.3). That is, Ineffable is not alone, as illustrated by Marcus' employment of several dyadic Valentinian syzygies and Tetrads, including those of pseudo-Valentinus and Epiphanes, which have dyadic tendencies. How he relates the Monad to Dyad is not specified, although he ascribes the engendering of all things not to one but to both (*Against Heresies* 1.16.1). Thus he combines both monadic and dyadic motifs, and shows no concern for reconciling them.[31]

Marcus adds considerably to the Valentinian repertoire of number symbolism. One of the most important new numbers in Marcus' system is six, which provides the bridge between the Ogdoad and the twenty-four–letter alphabet. Marcus notes six intrinsic components of each letter, and calls attention to the length of the name Ἰησοῦς. The Transfiguration is described with many references to six, and special attention is paid to the ἐπίσημον, the numeral six. Akhamoth is surprisingly described as six, not eight as might be expected from other Valentinian systems that make her an image of the Ogdoad.[32] Six is key, both to the creation (in conjunction with the Tetrad) of the twenty-four letters of the alphabet, and to the gap between the number of letters and the number of aeons. The number six also serves to introduce astrological symbolism, especially through Truth, whose twelve body parts, each assigned to two letters, are reminiscent of ancient astrological texts.

In many Valentinian systems the Ogdoad, a natural outcome of the unfolding of the Dyad, makes the number eight significant. It superficially resembled the ogdoad known from the ancient Egyptian cult at Hermopolis, where four male

---

[30] See also Förster 1999:306–310, on possible metaphysical parallels with late antique philosophy.
[31] See Förster 1999:301–302.
[32] But she is thought of as engendering the number seven, which is assigned, true to other Valentinian systems, to the Demiurge (*Against Heresies* 1.14.7).

deities and their female consorts were governed by the god Thoth.[33] But Marcus obsesses over the number. There are the eight semivowels, the eight letters in the name 'Christ,' the 888 in the name 'Jesus,' and the eight numerals for each of the digits, tens, and hundreds. Marcus' interest is particularly Christian, insofar as eight was a ubiquitous symbol in early Christianity (orthodox or not). For example, Sunday was commemorated as the eighth day, and some of the earliest baptismal fonts were octagonal.[34] Before Christianity eight was not popular, either with Pythagoreans or with other Greeks who used number symbolism, as is evident in the late antique dictionary of number symbolism, *The Theology of Arithmetic*, where the entry for the number eight is the shortest. Marcus is a very early witness to a new Christian claim on the symbolism of the number eight.

Another element to note is Marcus' interest in isopsephy. We have already seen in Irenaeus' extended system how the numeric value of the iota, the first letter in 'Jesus,' held important symbolism. This is similar to pseudo-Barnabas' exegesis of τιη´ (318) as a prophecy of Christ. But Marcus applies isopsephy to entire words, not just initials. The word περιστερά, because it adds up to 801, symbolizes Jesus (1.14.6, 1.15.1). And the word ἀμήν has the auspicious psephic value 99. The *Revelation to Marcus* is the first datable Christian attempt to use the practice not as a riddling device but as a tool for interpreting the Bible and language.

Did that tool verge on the world of magic and prognostication? We have seen elements in the *Revelation to Marcus* that would resonate with such practices: the zodiac-like Body of Truth and the primal utterance of the seven vowels. There is yet further evidence that Marcus was connected with prognostication. Hippolytus recounts at length a psephic technique that was used to determine which of two people would win in a contest of one sort or another (*Refutation of All Heresies* 4.14). One would take each contestant's name, add the numeric values of the letters, reduce each sum to a number between one and nine, then look up the two numbers on a chart, which would indicate the victor. According to Hippolytus, proponents of the technique developed numerous variations, a sign of its popularity. This form of numerology flourished in the later, Byzantine tradition, as confirmed by scores of manuscripts, which often credit the technique to Pythagoras. There is no evidence for psephic numerology before the

---

[33] Méautis 1918:20. The resemblance is only apparent in name, function, and even structure, since the Egyptian ogdoad was really an ennead. If in *Isis and Osiris* 3 Plutarch refers in veiled terms to Hermopolis' ogdoad, the enneadic structure is confirmed, since he makes Isis/Justice the head of the nine muses, i.e. the nine gods of Hermopolis. Plutarch's analogy, which does not correspond to the ancient ogdoad, is a reminder that non-Egyptians in late antiquity knew little about the ogdoad of Hermopolis. See Gwyn Griffiths 1970:264–265 and Méautis 1918:21, 24–25.

[34] For these and other parallels, see Quacquarelli 1973.

late second century, and ample evidence for it after.[35] Hippolytus implies that a certain Colarbasus was the inventor of this technique. And this very Colarbasus is said in the *Revelation* to be associated with Marcus as a fellow Valentinian (*Against Heresies* 1.14.1). Whether or not Marcus had a hand in numerology, and whether or not he collaborated with Colarbasus, he shows the same affinity for the intellectual apparatus that underlies numerical divination. Both Marcus and the numerological texts take psephy seriously, without relegating it to a parlor game or literary adornment. They both exhibit a common belief that psephy can reveal the hidden knowledge of the world. Whereas isopsephic prognostication delved into the mechanics of how to predict the future, Marcus explored a theology that could provide it with a conceptual foundation.

The *Revelation to Marcus* shows how adventurous and far-reaching Valentinian theology could be. The neo-Pythagorean philosophical and exegetical underpinnings have been smothered with layers of speculative arithmology. Marcus reaches into the far corners of culture and language, pushing the limits of Valentinianism and Christianity. As we shall see in the next chapter, he was not the only pioneer.

---

[35]   Luz 2010, an excellent introduction to the topic.

5

Alternate Paths in the
Early Christian Theology of Arithmetic

Monoïmus and the *Paraphrase of the "Apophasis Megale"*

A LTHOUGH THE VALENTINIANS PROVIDED A GREAT VARIETY of theologies of arithmetic, they were in reality part of a general trend. In this chapter
we turn to two unrelated systems, those of Monoïmus and the *Paraphrase of the
"Apophasis Megale,"* both attested almost exclusively in Hippolytus' *Refutation of
All Heresies*. The two systems overlap in some of their ideas, but they are distinct.
Each of them shares ideas with other systems Hippolytus discusses, and many
of these parallels, particularly those relating to protology and arithmology, will
be brought into my discussion as warranted.[1]

Monoïmus

All record of Monoïmus' existence would be lost, were it not for Hippolytus'
discussion and a fleeting mention by the fifth-century bishop Theodoret of
Cyrus, who says, in toto, "They say Monoïmus the Arab, getting his start from

---

[1] I sequester the so-called Naasenes, who open the fifth book of Hippolytus' treatise. Like Monoïmos, the Naasenes held to a "binitarian" system (Human–Son of Human; cf. *Refutation of All
Heresies* 5.6.4 with 8.12.2) and quoted from Orphic hymns (cf. 5.7.38 [plus refs. in the fontes of
Marcovich's edition] with 8.12.1). Like the *Paraphrase*, the Naasenes also used the *Apophasis Megale*
(cf. 5.9.5 with 6.9.4, 6.11.1, and 6.14.6), related fire to desire (cf. 5.8.16 with 6.17.4), deployed trees
and their fruit as a metaphor (cf. 5.8.31 with 6.9.9–10 and 6.16.6), and interpreted anthropologically the rivers of Eden (cf. 5.9.14–18 [plus refs.] with 6.14.8 and 6.15.1–6.16.4). All three groups
celebrate 'mother' and 'father' as paradoxical epithets of a single divine entity (cf. 5.6.4–5 [plus
refs.] with 6.17.3 and 8.12.5). In a pair of articles that are still profitably read (1882, 1887), Salmon
argued for a school of Ophites, of which Monoïmos was a part. This view of the makeup of the
Orphites has been altered, but not settled, by Frickel 1968 and Rasimus 2005. The Naasenes, as
well as the Orphites, who follow them in Book 5, diverge substantially from the two systems discussed in this chapter in metaphysics and number symbolism. An investigation of those aspects,
beyond the individual parallels discussed below, should be reserved for a study where they can
be examined and treated as a whole.

arithmetical knowledge, put together his own heresy," by far the shortest entry in his encyclopedia of heresies.[2] Hippolytus also says that Monoïmus was an "Arab" (possibly this means Syrian) but offers no other biographical information. The name Monoïmus is unattested in Greek literature, papyri, or inscriptions, although it may be a variation on the slightly less rare Monimos, related to the common Arabic name Munʿim, or its diminutive, Munayʿim.[3] The terminology he uses in his one extant work, his letter to Theophrastus, suggests that he had a respectable Greco-Roman education, which would put him in the wealthier class of early third-century society.

Hippolytus discusses Monoïmus' system in a terse ninety-one line passage of his *Refutation* (8.12–15). A summary of that system is followed by an alleged letter from Monoïmus to a certain Theophrastus (8.15), perhaps a cover letter attached to the text Hippolytus used for his summary. Thanks to that summary we know that Monoïmus posited two principles, which he called primarily ἄνθρωπος and υἱὸς ἀνθρώπου, referred to in this chapter as 'Human' and 'Son of Human,' awkward epithets that at least convey their applicability to both genders.[4] For Monoïmus the metaphysical core of the world is a vertical pole, with the higher principle, Human, blending into the lower one, Son of Human (see Figure 5). Human is unbegotten, incorruptible, and eternal, whereas Son of Human is begotten, subject to passion, and generated without time, will, or prior determination. Human is to Son of Human as being is to becoming, a Platonic analogy that explains why—so Monoïmus argues—some passages of Scripture distinguish between ἦν 'it was' and ἐγένετο 'it became', presumably because Scripture also holds to these two principles.[5] Monoïmus extends the analogy further. Human is to Son of Human as fire is to light, since light is generated concurrent with the fire's existence, without time, will, or prior determination (8.12.4). Monoïmus calls Human the 'one Monad' (μία μονάς), which he describes with a series of paradoxes: he is incompositely composite, indivisibly divisible, friendly and combative to all, peaceful and belligerent to all, dissimilarly similar, and like a kind of musical harmony (8.12.5). The appeal to harmony is telling of the thrust of Monoïmus' metaphysics, because in antiquity harmony was a kind of paradox, or even contradiction, since in it dissimilar tones were united. Thus Monoïmus' highest entity is at once a simple unity and a repository of plurality.

[2]   *Compendium of Heretical Fables* 1.18 (PG 83.369B).
[3]   Dillon summarizes the evidence for the name Monimos (1987:865). I thank Irfan Shahid for the suggestion of the Arabic name. The earliest attestations of Munʿim are most frequently Safaitic. See Harding 1971:569, s.v. "MNʿM."
[4]   On *Man/Human* as a title for a deity, see Dillon 1992:106–107 and Schenke 1962.
[5]   Genesis 1.2–3; John 1.1–4, 6, 9–10. See below for further discussion of Monoïmus' appropriation of Plato's *Timaeus*.

Figure 5. Depiction of the iota of Monoïmus. (Illustration by author.)

Monoïmus says that Human subsumes in himself all things, including contradictions and opposites. Three times he uses of Human the phrase μία μονάς, to describe Human's ability to transcend contradiction (8.12.5.15, 8.12.7.24–25, 8.13.1.1). The epithet, like the metaphors, is a paradox. It employs number symbolism to illustrate how Human unites and reconciles two incommensurate realms. As shown by the Tetrads of Epiphanes and Marcus (and other evidence discussed in Excursus A), the terms μονάς and ἕν were frequently differentiated, with the μονάς residing in a plane metaphysically higher than the ἕν. The former, an ideal entity, usually generates the latter, instantiated through physical, countable objects. The existence of the ἕν, of course, depends upon that of the μονάς. Monoïmus' term μία μονάς, then, draws upon this distinction (μία is the feminine form of ἕν). To the philosophically attuned, the term was as contradictory as 'thought thinker' or 'becoming being,' since it suggested the confluence of creator and creation, normally irreconcilable.

Monoïmus uses other paradoxes drawn from number symbolism and philosophical number theory. For instance, Human is both mother and father, and is described with both masculine and feminine nouns. This is the first system we have encountered that corresponds explicitly to the second of three Valentinian positions Irenaeus outlines concerning the Monad, namely that the Monad is androgynous, its own consort. Its androgyny extends to its role as Monad. As the source of numbers, and not a number proper, the Monad contains in itself potentially both odd and even, and therefore, by extension, female and male. This paradox is expressed best by the Greek letter *iota*, Monoïmus' preferred symbol for Human, whose perfection is most evident in his identity as the one iota (ἰῶτα ἕν) or the one apex—'apex' here referring to the serif atop the *iota* (μία κεραία; 8.12.6, 8.13.1). The language comes from the Sermon on the Mount (Matthew 5.18; see also Luke 16.17). Jesus' distinction in the Gospel accounts between 'iota' and 'apex'—the jot and the tittle—is to Monoïmus a veiled reference to a metaphysical structure. Like any good Christian exegete from this period, Monoïmus pursues the ramifications of that distinction, and he opts for the allographic component. The shape of the letter ἰῶτα reveals the relationship of Human to Son of Human (Figure 5). The iota and the apex are distinct but intertwined. The iota "is incomposite and simple" and yet is also composite and consists of many forms, shapes, and parts (8.12.6–7). "That single undivided object is the many-faced, myriad-eyed, and myriad-named single apex of the iota." And that apex "is the image [εἰκών] of the perfect Human."[6] For the analogy Monoïmus draws from the language of Paul. The apex as "the image of the perfect Human" alludes to Paul's description in Colossians 1.15: "The Son is the image of the unseen God."[7] In like manner, the iota and the apex are separate but interlocked. The former encompasses the latter, and the latter is the image of the former, language also used by the *Tripartate Tractate*.[8] Human is to Son of Human as the iota is to its apex.

Up to this point Monoïmus has not declared whether he is interested in the iota *qua* letter, or *qua* numeral. He now brings its numerical symbolism to the fore. He says that there is the one Monad, then the one apex, then the Decad of the one apex (8.13.1). He explains that this new entity, the Decad, is the power in the apex.[9] Then come the rest of the numbers, extending from the Monad: Dyad,

---

[6] *Refutation of All Heresies* 8.12.6–7: ὑποδείματος δὲ χάριν, τοῦ τελείου ἀνθρώπου <τούτου> κατανόει, φησί, <τὴν> μεγίστην εἰκόνα <ὡς> "ἰῶτα ἕν," τὴν "μίαν κεραίαν"· ἥτις ἐστὶ [κεραία μία] ἀσύνθετος, ἁπλῆ, μονὰς εἰλικρινής, ἐξ οὐδενὸς ὅλως τὴν σύνθεσιν ἔχουσα· <καὶ αὖ> συνθετή, πολυειδής, πολυσχιδής, πολυμερής. ἡ ἀμερὴς ἐκείνη μία <μονάς>, φησίν, ἐστὶν ἡ πολυπρόσωπος καὶ μυριόμματος καὶ μυριώνυμος μία τοῦ ἰῶτα κεραία, ἥτις ἐστὶν εἰκὼν τοῦ τελείου ἀνθρώπου ἐκείνου, τοῦ ἀοράτου.

[7] Monoïmus: ἥτις ἐστὶν εἰκὼν τοῦ τελείου ἀνθρώπου ἐκείνου, τοῦ ἀοράτου. Paul: ὅς [= υἱός, 1.13] ἐστιν εἰκὼν τοῦ θεοῦ τοῦ ἀοράτου.

[8] E.g. NH 1.5:116.28.

[9] I accept the δύναμις γὰρ αὐτῇ τὸ ῑ of the manuscript at *Refutation of All Heresies* 8.13.1.2 against Marcovich's reading δύναμις γὰρ αὕτη το‹ῦ› ἰῶτα. The term ἰῶτα implies a letter, not a numeral. Marcovich's reading is hard to reconcile with the normal senses of δύναμις and the arithmological discussion. ῑ refers to the numeral, more suitable to the context, not that this completely clarifies a convoluted passage. See also n. 10 below.

Triad, Tetrad, and so on, up to the Ennead and ten.[10] These are the complex (πολυσχιδεῖς) numbers, which reside in the simple and incomposite "one apex" of the iota. This explains Colossians 1.19 and 2.9, that "all the Pleroma was pleased to dwell bodily" in the Son of Human. For Monoïmus, these "sorts of combinations of numbers" become the bodily instantiations generated from the simple and incomposite "one apex" of the iota.[11]

Here numerical and allographical symbolism are combined into one metaphor. The prime image is that of the iota. It is a single character, created by a single stroke, yet its uppermost part, its serif, represents and mediates the whole and becomes the root of plurality. Monoïmus reinforces this analogy with the prevailing theories on the generation of numbers, where the first nine numbers, building blocks for all subsequent numbers, reside potentially in the monad. There is a parallel here with Nicomachus, who describes the cosmos as "rooted" in the monad, but made and revealed in the Decad.[12] So too in Monoïmus' view, the Son of Human, as the ἰῶτα ἕν, a being that synthesizes ten and one, is the source, completion, and regulator of creation (8.13.4.21).

Monoïmus interprets the hexaemeron of Genesis in light of his decadology. The six days of Creation are six powers trapped in the one apex of the iota. The Sabbath comes into existence from the Hebdomad of the world beyond (ἀπὸ τῆς Ἑβδομάδος γέγονε τῆς ἐκεῖ), probably referring by the term 'Hebdomad' to the iota itself, combined with the six powers of Creation. That is, the iota-Human sends forth a seventh power, which is represented by the Sabbath. These powers are the source of the four material elements, from which the cosmos is made. Thus the seven powers are a connective tissue between Human and the material universe. The emanation of six or seven powers from a single power has parallels elsewhere. In one of his elusive metaphysical models, Philo posits six latent powers.[13] Hippolytus' Valentinians have six powers organized into syzygies and

---

[10] *Refutation of All Heresies* (Marcovich) 8.13.1: Ἔστιν οὖν, φησίν, ἡ ‹μία› μονάς, ἡ μία κεραία, καὶ δεκάς· δύναμις γὰρ αὕτη το‹ῦ› ἰῶτα, τῆς μιᾶς κεραίας, "ἐν ᾗ ἐστιν ἡ παντὸς ἀριθμοῦ ὑπόστασις· μονὰς" καὶ δυὰς καὶ τριὰς καὶ τετρὰς καὶ πεντὰς καὶ ἑξὰς καὶ ἑπτὰς (καὶ) ὀγδοὰς καὶ ἐν‹νε›ὰς μέχρι τῶν δέκα. Marcovich's third emendation (underlined) is excessive for three reasons. It breaks up the chain of entities (one Monad, one apex, Decad, Dyad, Triad, Tetrad, etc.), it suggests that the Monad begets the Monad, and it classifies the Monad as a complex (πολυσχιδής, line 5) number. By my reconstruction of Monoïmus' system, the text of *Refutation of All Heresies* 8.13.1.1–4 should read: Ἔστιν οὖν, φησίν, ἡ ‹μία› μονάς, ἡ μία κεραία, καὶ δεκάς (δύναμις γὰρ αὐτῇ τὸ ῑ) τῆς μιᾶς κεραίας, καὶ δυὰς καὶ τριὰς καὶ τετρὰς καὶ πεντὰς καὶ ἑξὰς καὶ ἑπτὰς (καὶ) ὀγδοὰς καὶ ἐν‹νε›ὰς μέχρι τῶν δέκα.

[11] αἱ γὰρ τοιαῦται τῶν ἀριθμῶν συνθέσεις, reading τοιαῦται of the manuscript against Wendland's τοσαῦται (1916).

[12] *Theology of Arithmetic* in Photius *Bibliotheque* 187:144a25–27.

[13] See Stead 1969:80, depending upon *On Flight and Finding* 95–96. But that passage could be understood to mean one prime power that governs five subpowers.

governed by a Monad. For seven powers, there is the more remote parallel at a temple at Esna, Egypt, where seven gods exit the mouth of a single goddess.[14] These earlier systems do not completely explain Monoïmus' protology. Rather, they show that his numerical theology, and therefore the biblical exegesis upon which it depended, was the result of a shared vocabulary, not direct copying.

Monoïmus links his cosmology not only to the account in Genesis but to the *Timaeus*, one of the most influential of Plato's texts in late antiquity. In that work, all the material elements are thought to be composed of two kinds of triangles, isosceles and equilateral. The triangles adhere to each other to forge the world's five basic molecules, only four of which Plato discusses: cubes, four-sided pyramids, octahedra, and icosahedron. These correspond to earth, fire, air, and water.[15] Monoïmus concurs with Plato's scheme, and claims that the isometrical shapes themselves come from the numbers retained by the apex of the iota (8.14.1–2). His logic is that the numbers behind the five geometrical figures of the *Timaeus* must have some source or origin. Monoïmus identifies that source as the apex of the iota, the Son of Human. This is an appeal not merely to understand Moses in terms of the *Timaeus* but to read Plato in light of Monoïmus.

The symbolism behind the allograph ι comes into play in his biblical exegesis, just as it does in his metaphysics. Monoïmus considers it significant that Moses uses his rod to generate exactly ten plagues. The shape of the rod in its variegated simplicity represents the iota and its apex. Because the iota resembles the fruitfulness of a vine, he says, it reflects the creation of the world (8.14.3). Monoïmus relates the striking action latent in the term *Decaplague* (δεκάπληγος)—Moses' ten plagues, or "blows"—to the severing of the umbilical cord at birth. Ultimately, both are conducive to generation (8.13.4, citing Democritus, frag. 32). Indeed, Monoïmus later claims that the transformation of creation is actualized (ἐνεργεῖται) by the Decaplague (8.14.8). The Decalogue and the Pentateuch, too, are each derived from the numbers resident in the one apex. The Decalogue, just as the Decaplague, is based on the Decad, a portal for knowing the universe. The Pentateuch derives from the Pentad, also kept in the one apex (8.14.5).

The Decad is also central to Monoïmus' interpretation of the dates of the Jewish Pascha.[16] He claims that the fourteenth day of the month is the source (ἀρχή) or origin of the Decad. It may seem puzzling to think that fourteen is the source of ten, since the former comes after the latter. Monoïmus' explanation?

---

[14] Förster 1999:185–186.
[15] Plato *Timaeus* 55a–56b.
[16] See also p. 131 below.

The numbers one through four add up to ten, which is the perfect number and the one apex.[17] The process of deriving ten from four is symbolized by the number fourteen, ιδ´, the digits of which represent the first four numbers and their total sum. Fourteen as the source of ten is intelligible only in light of Monoïmus' allographic symbolism, where ιδ´ represents the one apex unfolding from potential decad (δ´ being shorthand for the Pythagorean τετρακτύς) into the actual decad of the stem of the iota. This is the source of ten (8.14.6). The festal dates also point to the seven powers of creation. The Hebdomad derived from observing the festival from the fourteenth to the twenty-first days is itself the creation of the world, which also resides in the one apex.[18] Thus, by virtue of the numbers embedded in their dates of celebration, Pascha and the Feast of Unleavened Bread represent the causes of creation;[19] and the Decaplague, its transformation and change (8.14.8).

Hippolytus accuses Monoïmus, among other things, of reading Moses in terms of Greek wisdom, specifically of using Aristotle's ten categories to interpret the Law (8.14.9). This echoes another passage in the *Refutation* (6.24) where he ascribes to Pythagoras a system that looks suspiciously like that of Monoïmus. This earlier passage is part of a longer exposé (6.23–28) of doctrines attributed, by Hippolytus or his source text(s), to Pythagoras. One scholar suggested that "a Gnosticizing Pythagorean treatise" was Hippolytus' direct source for the longer exposé.[20] But some sections of 6.23–28 have no recognizably gnosticizing inclination, whereas others do; some sections have nothing in common with neighboring passages, aside from an affinity for neo-Pythagorean ideas and texts. And Hippolytus alternately attributes the ideas to Pythagoras, the Pythagoreans, and unnamed individuals. Thus I regard 6.23–28 as a potpourri, dressed by Hippolytus or an earlier anthologist to look like a single account.

At 6.24, the passage of central interest to us here, "Pythagoras" differentiates between two worlds, the noetic and the sense-perceptible. The noetic world has as its source the Monad, whereas the source of the sense-perceptible world is the τετρακτύς, which possesses the iota, the one apex, and a perfect number. According to the Pythagoreans this ten (literally, ῑ) is the one apex, which is the first and foremost essence (οὐσία, Aristotle's first category) of noetic things.[21] These references to the iota and the one apex as a source of generation suggest that at 6.24, Hippolytus relies for his recreation of "Pythagoras'" teaching upon

---

[17] On ten as a perfect number, see p. 54n78 above.

[18] Exodus 12.15–20.

[19] Or the elements (στ<οιχ>εῖα) of creation, if we adopt Marcovich's emendation of Hippolytus' *Refutation of All Heresies* 8.14.8.38.

[20] Marcovich 1986:23.

[21] And of sense-perceptible things, if Wendland's emendation at 6.24.1.5 is correct (1916).

a text written by Monoïmus or someone in the same circle. (For convenience, I refer to the author of that text herein as 'Pythagoïmus.')[22]

Like Monoïmus, Pythagoïmus argues that there are nine accidents that occur in οὐσία, and he proceeds to list the remaining nine (Aristotelian) categories (6.24.2).[23] The total, he claims, possesses the perfect number, ten. This comment accords with the Pythagorean tendency to claim Aristotle's ten categories for their own tradition, usually under the name Archytas, and to explore their numerical symbolism.[24] There may be something substantial, then, behind Hippolytus' complaint that Monoïmus read the Law in terms of the ten categories, which he lists in full as if Monoïmus had discussed them seriatim (8.14.9). Given Pythagoïmus' interest in the ten categories and their connection to realms of mental and sense perception, Monoïmus too may have interpreted the Pentateuch in light of Aristotle's categories in a treatise no longer extant. An author such as Monoïmus would have been interested in exploring one-to-one correspondences between the categories, the Ten Commandments, and the ten plagues.

Monoïmus' letter to a certain Theophrastus, with which Hippolytus concludes his entry on Monoïmus, suggests further comparisons with Pythagoïmus' number symbolism. The letter starts off by admonishing Theophrastus that if he wishes to know God, he should stop looking for him in creation, but rather look for God within himself, "and learn who it is that appropriates to himself absolutely everything in you" (8.15.1). To this end Monoïmus advises him to say, "My God, my mind, my understanding, my soul, and my body."[25] The five items are listed in descending hierarchical order, resonating well with Monoïmus' arithmological interest in the number five (8.14.5), especially as it relates to the Pentateuch and anthropology, a theme found in other authors, including Pythagoïmus.[26] Monoïmus promises that one who follows his advice of intro-

---

22 The author need not be Monoïmus himself. At *Refutation of All Heresies* 8.14.9, Hippolytus summarizes the preceding summary of Monoïmus' doctrine by referring to "these men," which suggests a circle in which Monoïmus traveled.

23 Commentators in late antiquity took the order of the categories seriously; see Chiaradonna 2009. The list of categories at 6.24.2 is identical to that at 8.14.9 (although see the note at line 49 in Marcovich's edition [1986:335]), but different from that at 1.20.1 in the placement of quality, quantity, position, and state. The first two passages, by Monoïmus and Pythagoïmus respectively, further corroborate the closeness of the two authors, and their possible reliance on Eudorus, who may have been the first to arrange the categories in this order. See Dillon, who notes that Philo consciously uses a different order of Aristotle's categories (1996:134–135 and 178–180).

24 See e.g. pseudo-Archytas' *Ten Universal Categories*, ed. Thesleff 1965:3–8.

25 [ὁ θεός μου,] ὁ νοῦς μου, ἡ διάνοιά μου, ἡ ψυχή μου, τὸ σῶμά μου· Marcovich excludes the phrase "my God," but against the manuscript and at the expense of Monoïmus' metaphysical hierarchy, as explained here.

26 Pythagoïmus, at *Refutation of All Heresies* 6.24.3–4, on the five senses; Clement of Alexandria presents man as possessing a decalogue of faculties, composed of two quintets (*Stromateis* 6.134.2).

spection and accurately diagnoses his own emotions and motivation will eventually discover God, who is both "one and many according to that one apex," and find escape from himself (8.15.2).[27] If the process of knowing God follows the sequence God → mind → understanding → soul → body, then in theory the escape would occur along the same sequence, in reverse. The five-stage process from material corruption to divine reality is guided by the one apex.

The arithmology of Monoïmus is rather distinct from the Valentinian. The Pleroma, which dwells in the Son of Human, is made up not of aeons but of numbers, essentially the first Decad. Missing from his arithmology is any reliance upon syzygies or eights, both of which are centrally prominent in Valentinianism. Instead, the seven powers that emanate from the iota are undifferentiated and unnamed (or Hippolytus has uncharacteristically omitted this detail). Monoïmus' numbers are channeled into the cube, icosahedron, octahedron, and pyramid—a model of creation inspired by the *Timaeus*—so as to produce the material world. This contrasts with Valentinian cosmologies, which were less committed to the *Timaeus*. Another difference is that Monoïmus' numbers are agents of transformation, a process pleasing to God since it helps rescue people from deception. In Valentinianism, numbers do not play this active a role in salvation. Finally, the sparse protology of Monoïmus, who propounds only two "aeons"—Human and Son of Human—resembles the simpler, earlier forms of Valentinianism, witnessed in the *Gospel of Truth* and the *Tripartate Tractate*. The names Human and Son of Human, one derived from the other, match this metaphysical simplicity.[28] Accordingly, Monoïmus can be described as being monadic, not dyadic, in his protology. But his Monad is androgynous and therefore in no need of a consort. Monoïmus' pure monism serves the purposes of Hippolytus, since it lets him place both Monoïmus and Pythagoïmus in the line of Pythagoras, who, Hippolytus is convinced, was the arch-monist.

Monoïmus' use of interdependent number and letter symbolism suggests an affinity with Marcus. But Marcus, despite the complexity of his *Revelation*, does not show the sublimity seen in Monoïmus' simple but thoughtful identification of the iota and its apex with Human and Son of Human. Monoïmus, unlike Marcus, explores the theological significance of an allograph, the ι, his preferred way to illustrate the nature of God. He was an unusual theologian who tried to fuse in the physical symbol of the letter and numeral ι a range

---

See pp. 101 and 128–136 below.

[27] εὑρήσεις αὐτὸν ἐν <σ>εαυτῷ, ἓν <ὄντα> καὶ πολλά, κατὰ τὴν κεραίαν ἐκείνην <τὴν μίαν>.

[28] Monoïmus' scheme, or a related one, seems to have crept into Epiphanius' account of Colarbasus (*Panarion* 35.2.4–12), whom he accuses of giving to the Father the name *Human*, on the basis of the Savior's saying he was the Son of Human. None of the rest of Epiphanius' discussion of Colarbasus can be attributed to Monoïmus (or to Colarbasus, for that matter), since the system it describes is an advanced variation of Valentinianism.

of philosophical and religious perspectives: a simple, "binitarian"-like theology, Moses' account of creation, Plato's cosmogony, and Aristotle's categories. In his attempt to collapse all of philosophy and scripture into a single letter-*cum*-numeral, Monoïmus is unique.

The *Paraphrase of the "Apophasis Megale"*

According to Acts 8, Philip traveled to the city of Samaria, to preach and perform miracles. Resident there was a certain Simon, a magician who enjoyed local popularity, evident in the title the entire town gave him: the Great Power of God. After a number of his admirers were baptized, Simon too believed, was baptized, and watched with admiration first Philip and then Peter and John perform miracles and bestow the gift of the Holy Spirit. Simon tried to buy this power. When Peter rebuked him and exhorted him to repent, Simon meekly asked for mercy. The Bible's silence on whether or not Simon truly repented inspired numerous colorful post–New Testament accounts about Simon. Some made him the father of all heresies that flourished in the second century; others made him a hero and claimed to preserve his teachings; but none preserved a tradition that can be reliably traced back to him.[29]

Although Hippolytus was in the anti-Simon camp, his heresiology preserves substantial parts of the early Simonian tradition, including texts that are known nowhere else. Several times he quotes Simon himself from a text he calls either the *Great Revelation* (Ἀπόφασις μεγάλη) or merely the *Revelation*, commonly referred to today as the *Apophasis Megale*. When Hippolytus' text was first edited and published, in the mid-nineteenth century, the quotations from Simon generated interest among New Testament scholars attempting to reconstruct the historical Simon. The hopes that his words might be preserved in Hippolytus' text were dampened when Frickel demonstrated that Hippolytus was citing not the original *Apophasis Megale* but an anonymous paraphrase of it, composed in the early third century.[30] Since then, most studies of the *Paraphrase of the "Apophasis Megale"* have continued to mine it for the original, first-century *Apophasis Megale*. The paraphrase tends to be of little interest since it postdates the original Simon by more than a century.[31] But the *Paraphrase*, and not the *Apophasis Megale*, is important for this study for its creative use of number symbolism.

The *Paraphrase of the "Apophasis Megale"* purports to be "the book of revelation of *Phone* and *Onoma* from the *Epinoia* of the great Power, the unbounded"

---

[29] The best full-length study on Simon Magus is Heintz 1997. More recent but less helpful, in part because it does not interact with Heintz's study, is Haar 2003.

[30] Frickel 1968.

[31] Exceptions to this tendency are Mansfeld 1992:166–177 and Edwards 1997.

(*Refutation of All Heresies* 6.9.4).[32] This, the opening line, suggests the beginning of an apocalyptic or revelatory text. The *Paraphrase* is just that, but it is also a commentary, both on the Bible and the *Apophasis Megale*. And it is a metaphysical treatise, very similar to that of Monoïmus. The author of the *Paraphrase* has a vision of how the highest tiers of the world are structured, and he uses disparate and seemingly unrelated texts to describe that universe.

As reported by Hippolytus, the author of the *Paraphrase* (whom I call for convenience "deutero-Simon") considers the root of the universe to be the Infinite Power (ἀπέραντος δύναμις, 6.9.5). This title, repeated twenty times in Hippolytus' account, is important to deutero-Simon, who seems to have been the first to use it.[33] It is a subtle polemic against the New Testament, where Simon is called the Great Power.[34] 'Infinite' is infinitely greater than 'Great,' and his preference for the former over the latter suggests that above Simon, the great power, was a higher power to which he was subordinate, thus answering accusations that Simon thought he was God.[35]

The name also toys with an important part of the ancient Pythagorean tradition. Ἀπέραντος and ἀόριστος (and cognates)—'infinite' and 'limitless'— were traditionally applied to the Dyad, not to the Monad.[36] Rather, the Monad was thought of as a limiting force, an agent that brought stability and shape to an unformed Dyad. Hence Plato's Indefinite Dyad. For deutero-Simon the opposite is true: it is not the Dyad that is "without bound," but the one greatest power (ἀπέραντος δύναμις). In contrast, Thought, Name, and Consideration, the second, fourth, and sixth powers (discussed below), all complete or limit their conjugal counterparts. Thus, in the *Paraphrase* the dyadic, female powers limit the odd powers, the reverse of the classic Pythagorean and Platonist model.

The Infinite Power, which has a twofold nature, is the metaphysical foundation of the system. It consists of two aspects, hidden and visible (6.9.5). Fire, the most fundamental element in the universe, is an example. It does not have, as many think, a single, simple nature. Rather, its nature is twofold: its hidden aspect hides in its visible one, and the visible aspect is brought into existence

---

[32] τοῦτο τὸ γράμμα Ἀποφάσεως φωνῆς καὶ ὀνόματος ἐξ ἐπινοίας τῆς μεγάλης δυνάμεως τῆς ἀπεράντου; trans. Mansfeld 1992:173n56.

[33] Cf. the undatable Hermetic fragment 28, cited by Cyril of Alexandria *Against Julian* 1.46.10–19, although far from deutero-Simon's near-technical use. See also Hermetic fragment 26. Hippolytus' terminology is not everywhere consistent. At *Refutation of All Heresies* 6.13.1.13 and 6.18.3.10, the Infinite Power is called the Great Power.

[34] Acts 8.10: Οὗτός ἐστιν ἡ δύναμις τοῦ θεοῦ ἡ καλουμένη Μεγάλη. Literally, "the power of God, so-called Great."

[35] The very charge Hippolytus makes at 6.14.1.

[36] See e.g. Aristotle *Metaphysics* 1081–1083, where the ἀόριστος δυάς is discussed. In the Pythagorean tradition influenced by Philolaus, it is chiefly "infinites" or "unlimiteds" (ἄπειρα) that correspond to the dyad and to even numbers.

by the hidden one (6.9.5–6). Deutero-Simon links this polarity to the distinction Aristotle makes between potentiality and actuality, and Plato's contrast between mental and sense-perceptible objects. Throughout the *Paraphrase* such bifurcation recurs, in terms drawn not only from sense perception (visible versus invisible, audible versus the voice itself) but from arithmetic, from the distinction frequently made between numbers and numerable things (6.11.1).[37] The Infinite Power bestows this bipartite structure on man by creating him "in the image and in the likeness," a verse that deutero-Simon interprets in light of bipartite nature, assigning to "image" and the upper part of man the spirit who "hovers over the waters" (6.14.5–6, citing Genesis 1.2, 26). Hippolytus does not mention what deutero-Simon assigns to the lower half, the "likeness"; presumably this is the soul.

The Infinite Power is called the root of the universe (6.9.5, 6.17.3). Deutero-Simon develops the idea further, by likening the Power to the tree seen by Nebuchadnezzar (6.9.8; Daniel 4:10–12).[38] The trunk, branches, and foliage are the visible half. The purpose of the tree is to produce perfect, well-shaped fruit, which unlike the other visible elements of the tree will be put into the storehouse rather than into the fire (6.9.8–10).[39] The fire that consumes the tree

---

[37] For my distinction "between numbers and numerable things" I depart from Marcovich's Greek text, which describes the two parts of the fire: Τοιούτου δὲ ὄντος, ὡς δι᾽ ὀλίγων εἰπεῖν, κατὰ τὸν Σίμωνα τοῦ πυρός, καὶ πάντων τῶν ‹μερῶν αὐτοῦ,› ὄντων ὁρατῶν καὶ ἀοράτων, ἐνήχων καὶ ‹ἀν›ήχων, ἀριθμητῶν καὶ ‹ἀν›αρίθμων, ‹φρόνησιν ἐχόντων›—ὧν αὐτὸς ἐν τῇ Ἀποφάσει τῇ μεγάλῃ καλεῖ τελείων νοερῶν—, [οὕτως ὡς] ἕκαστον τῶν ἀπειρά(κι)ς ἀπείρων ‹μερῶν ἐπιδέχεται› ἐπινοηθῆναι ‹ὡς› δυνάμενον καὶ λαλεῖν καὶ διανοεῖσθαι καὶ ἐνεργεῖν, οὕτως ὡς, φησίν, Ἐμπεδοκλῆς ‹λέγει› (6.11.1). Part of the problem with Marcovich's restoration (which follows Wendland 1916) is that ἄνηχος, although sensible as an antonym of ἔνηχος, is unattested in Greek literature. His rendition of the third pair, ἀριθμητῶν and ‹ἀν›αρίθμων, does not create the clear-cut opposites he seems to intend, evident in their translation "countables and innumerables": one could have countable things that are too numerous to count, such as grains of sand. The manuscript without emendation—ἀριθμητῶν καὶ ἀρίθμων—makes sense to me. It draws from an idea popularized by Moderatus of Gades (fragment 2) and Theon of Smyrna (*Mathematics Useful for Reading Plato* 19.18–20.2), and further attested in Plotinus *Ennead* 6.6.9: numbers constitute a metaphysical order higher than countable things. For Theon, the monad is to numbers as the ἕν is to countables. See Excursus A. Note, too, that deutero-Simon's first pair is a contrast not so much of opposites as of metaphysical superior and dependent, illustrated in the analogy of fire at 6.9.6. At 6.11.1 is a list not of opposites but of correlative, hierarchical pairs. Thus, ἐνήχων καὶ ἤχων needs no emendation. A voice, after all, can be treated as the metaphysical superior to things heard, and the relationship of voice to sound mirrors that of number to countable object.

[38] On "root" as a theological metaphor in gnosis, see Attridge and Pagels 1985:23.217–218.

[39] The LXX version of Daniel, unlike the Theodotian version (the latter supplanted the former in the ancient Church, unique to Daniel), emphasizes a single root's being left on the tree. Daniel 4.15 LXX: καὶ οὕτως εἶπε Ῥίζαν μίαν ἄφετε αὐτοῦ ἐν τῇ γῇ, ὅπως μετὰ τῶν θηρίων τῆς γῆς ἐν τοῖς ὄρεσι χόρτον ὡς βοῦς νέμηται. It seems that LXX Daniel formed the basis of deutero-Simon's interpretation, which starts with Nebuchadnezzar's tree, but grafts onto it the teachings found elsewhere (especially the early gospels: Matthew 3.10, 7.19; Luke 3.9; *Gospel of Philip* 123) concerning trees.

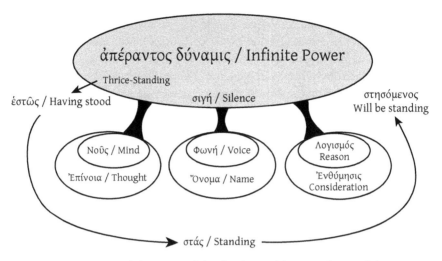

Figure 6. Partial depiction of the theology of the *Paraphrase of the "Apophasis Megale."* Arrows indicate actions; darkened stems, lines of generation. (Illustration by author.)

is the Infinite Power itself, which begets the cosmos and the first six roots of the beginning of creation (6.12.1). These roots emerge from the fire as three syzygies: Mind and Thought (Νοῦς, ᾽Επίνοια), Voice and Name (Φωνή, ῎Ονομα), Reason and Consideration (Λογισμός, ᾽Ενθύμησις; see Figure 6). According to deutero-Simon the entire Infinite Power resides in these six roots, potentially, not in actuality, and the six roots allow a person to unite with the Infinite Power in essence, power, size, and perfection (6.12.3). Should a person not make use of the six powers latent in the soul and be fully formed, that person is destroyed and perishes. According to deutero-Simon, some people in a similar manner ignore the grammatical or geometrical knowledge latent in their souls, much to their loss.

In addition to the Infinite Power and the six powers there is a seventh power, given a title consisting of three participial forms of ἵστημι 'I stand': ἑστὼς στὰς στησόμενος 'having stood', 'standing', 'will be standing'.[40] Each participle corresponds to one of three stages in this seventh power (which for convenience I call "Thrice-Standing"). In the "having stood" stage (ἑστώς), it resided above, in the unbegotten Power. In "standing" (στάς), it is begotten below in the flow of waters, in the image of the Infinite Power. It "will be standing" (στησόμενος) above, alongside the Infinite Power (6.17.1). This seventh power, then, unlike the other six, begins in the Infinite Power, sojourns in the lower world as an image

---

40 For textual parallels to this phrase in Hippolytus and other works see Marcovich 1986:214, note to line 5. See also M. Williams 1985.

of the Infinite Power, and then ends beside the Infinite Power. Throughout the *Paraphrase* the Thrice-Standing and its source, Infinite Power, are so closely identified they are sometimes indistinguishable. Nonetheless, they are distinct.[41]

The Thrice-Standing acts on behalf of the Infinite Power by descending to creation and waiting to be perfected in beings, so as to bring them back to the Infinite Power. The other six powers have the Infinite Power latent within them, but the seventh power, the Thrice-Standing, perfects their work by raising to the side of the Infinite Power persons who have been perfected (6.12.2). Because of its special mission, the Thrice-Standing is the subject of a number of cryptic or paradoxical epithets, and possibly even worship. The Thrice-Standing, deutero-Simon says, explains the saying, "I and you are one; before me, you; after you, I."[42] This single power is "divided up and down, begetting itself, growing itself, seeking itself, finding itself, being its own mother, its own father, its own sister, its own conjugal union, its own daughter, its own son" (6.17.3).[43] Paradoxes of this type are reserved in other systems normally for the One or the Monad, but here they describe a power that emanates from, and returns to, the Infinite Power.[44]

Toward the end of Hippolytus' account of the *Paraphrase of the "Apophasis Megale,"* deutero-Simon explains more fully—and cryptically—the internal structure of the three syzygies (6.18.2–7). He says that for all the aeons there are two "shoots" (παραφυάδες), a term that extends his tree analogy.[45] These shoots

---

[41] At 6.12.3.10, the Thrice-Standing seems to be conflated with the Infinite Power: εἶναι δὲ ἐν ταῖς ἐξ ῥίζαις ταύταις πᾶσαν ὁμοῦ τὴν ἀπέραντον δύναμιν δυνάμει, οὐκ ἐνεργείᾳ, ἥντινα ἀπέραντον δύναμιν ‹εἶναί› φησι τὸν ἑστῶτα ‹στάντα› στησόμενον. But this identification depends upon Marcovich's insertion of εἶναί and στάντα. Note that the text omits the second "standing" (στάντα), the one stage when the Thrice-Standing is away from the Infinite Power. The verb to be understood here is perhaps not εἶναι but ἔχειν, an emendation that would highlight the Thrice-Standing's two stages that are in the presence of the Infinite Power. At 6.14.2, however, deutero-Simon identifies the Infinite Power with the seventh power, which he calls the Thrice-Standing at 6.13.1.9. The account at 6.14.2 may depend upon a passage in the *Apophasis Megale* that calls the Thrice-Standing "infinite" by virtue of its special relationship to the Infinite Power. At any rate, 6.17.1 articulates the distinction: "having stood" (ἑστώς) is *in* the unbegotten power, "standing" (στάς) is in its image, and "will be standing" (στησόμενος) will be *alongside* the Infinite Power. At 6.14.3 is repeated the idea that the seventh power, another epithet for the Thrice-Standing, exists *in* the Infinite Power. On balance, then, the Thrice-Standing and the Infinite Power are distinct entities.

[42] 6.14.2: ἐγὼ καὶ σὺ ἕν· ‹τὸ› πρὸ ἐμοῦ σύ, τὸ μετὰ σὲ ἐγώ. See Marcovich 1986:222–223, note to line 10 for numerous close, but inexact, parallels in other ancient texts. See also Clement of Alexandria *Stromateis* 2.25.2.

[43] αὕτη, φησίν, ἐστὶ‹ν ἡ› δύναμις μία, διῃρημένη ‹δ'› ἄνω κάτω, αὑτὴν γεννῶσα, αὑτὴν αὔξουσα, αὑτὴν ζητοῦσα, αὑτὴν εὑρίσκουσα, αὑτῆς μήτηρ οὖσα, αὑτῆς πατήρ, αὑτῆς ἀδελφή, αὑτῆς σύζυγος, αὑτῆς θυγάτηρ, αὑτῆς υἱός, [μήτηρ, πατήρ,] ἓν οὖσα.

[44] Other early Christian traditions refer to a standing god or other entity, but these beings function somewhat differently from each other and from the *Paraphrase*. See M. Williams 1985:37–38, 57.

[45] I take as objective (and not subjective) the genitives in δύο εἰσὶ παραφυάδες τῶν ὅλων αἰώνων.

or branches come out of one root, the power called Silence.[46] In the first syzygy, the great power that is in the upper half is termed Mind and the bottom half is called Thought, the female part (6.18.3). The upper half is male and governs all things; the lower half is female and begets all things. The gap between the two is filled with "ungraspable air," and it has neither source nor boundary, wording that suggests that the gap is the Infinite Power itself.[47] The Father—the name here for the seventh power, the Thrice-Standing—nourishes in this gap all things that have source or boundary (6.18.4). This Father, just like the Infinite Power, is an androgynous power, and exists in the Monotes, from which Thought proceeds.

The *Paraphrase*'s account of the generation of the powers is confusing. It starts off with three entities—the one root with two shoots is the Father with Mind and Thought—but it then describes Thought as proceeding from the Father, as if Father and Mind were the same, and as if there were only two entities. Further, Silence's role in generating the syzygies is mentioned, but never explained. Despite this confusion, it is apparent that deutero-Simon considers the generation of the syzygies to be organic and internal to the Infinite Power. The male half of each syzygy is alone, although latently possessing the female part (6.18.5). He becomes "first" only after he generates the "second" through an act of self-introspection that reveals his Thought. The act is described as the Father's "issuing forth himself from himself," whereby he makes manifest his Thought.[48] This second figure now calls the first "Father," and hides him in herself, and the union creates an androgynous being: Power and Thought, with the Power as the upper half and the Thought as the lower (6.18.6; here δύναμις seems to be equated with Πατήρ or Νοῦς). This explains a phrase, presumably from the *Apophasis Megale*: "Being one, two are found" (ἓν ὂν δύο εὑρίσκεται). The male and female emerge from an androgynous, monadic model. The male initially carries the female latently, then the female emerges and surrounds the male.[49]

The numerically inspired doctrine of syzygies is central to the *Paraphrase*'s exegesis of Scripture, and it gives the number symbolism some prominence. Deutero-Simon brings to the creation account of Genesis his doctrine of the six powers and the Thrice-Standing. To begin, he assigns to each of the six powers key parts of creation: Mind and Thought are heaven and earth. Just as Mind oversees and guards his consort, Thought (who, in turn, receives his seed), so the masculine heaven looks down upon earth, which receives that which heaven sends

---

[46] Note now the late introduction of the classic Valentinian name for the second aeon.

[47] The Infinite Power has no limit (ἀπέραντος), and the gap has no limit (μήτε πέρας ἔχοντα): 6.18.3.13.

[48] 6.18.6: ὡς οὖν αὐτὸς ἑαυτὸν ἀπὸ ἑαυτοῦ προαγαγὼν ἐφανέρωσεν ἑαυτῷ τὴν ἰδίαν ἐπίνοιαν.

[49] Compare the Barbelo-Gnostic protology, p. 59 above.

down. Voice and Name are the sun and moon, and Reason and Consideration are the air and water (6.13). He then mentions the Thrice-Standing, calling it the seventh power, and thereby associates it with the seventh day, "the cause of the good things praised by Moses," who said "very good" (καλὰ λίαν; 1.13.1.10, citing Genesis 1.31 LXX). The only time "very good" is used in Genesis 1 is at the end of the sixth day, to summarize the creation. Days one through five are called merely "good." Deutero-Simon's point is that κατὰ λίαν distinguishes the six days of creation from the Sabbath, the day that perfects the goodness of the previous six. Likewise, the Thrice-Standing is the cause of goodness in the six powers. Deutero-Simon continues his extended comparison of creation and the powers, but with uneven results. He says that the three days that occur before the creation of the sun and moon (i.e. Voice and Name) refer to the first syzygy (Mind and Thought) and the seventh day of creation (6.14.2).[50]

Deutero-Simon applies number symbolism to other biblical texts as well. For instance, he differentiates the garden of paradise from Eden, in line with his tendency to identify binary pairs (6.14.7–8). The first, the garden of paradise, is a womb, as Isaiah says, "I am the one who fashioned you in the womb of your mother" (Isaiah 44.2, 24). The second, Eden, is a membrane, afterbirth, and navel, since it is "a river proceeding out of Eden to water paradise" (Genesis 2.10). The four springs that flow out of Eden resemble the four channels that are attached to the embryo. Two of these convey breath (or spirit, depending on how we take πνεῦμα) and two, blood.[51] The four rivers of Genesis further symbolize how the embryo has only four of the five senses—sight, hearing, smell, and taste. These four senses, in this, the classical sequence, are alluded to by the titles and content of each of the first four books of the Pentateuch. Genesis is sight, Exodus hearing, Leviticus smell, and Numbers taste (6.15.2–6.16.3).[52] The fifth sense, touch, is addressed by the title of the fifth book, Deuteronomy, which is geared to formed children, to confirm and summarize their other four senses.[53]

---

[50] It is unclear how the seventh Power, earlier assigned to the Sabbath, can now consistently represent one of the first three days of creation.
[51] See Pouderon 2005. Deutero-Simon integrates into his theology common beliefs about the fetus. In other parts of the *Paraphrase of the "Apophasis Megale"* the breath/spirit is seen as the higher aspect in ἄνθρωπος (Hippolytus *Refutation of All Heresies* 6.14.6), and the seventh power itself, as the image of the Infinite Power (6.14.4). The blood is fire, the sources of things begotten (6.17.4). Just as the root of all bifurcates into shoots (6.18.2–7, discussed above), so does the blood, into semen in men and milk in women (6.17.6).
[52] Compare *Refutation of All Heresies* 5.9.13–18, the parallel account of the Naasenes.
[53] At 6.15.1.3 the manuscript reads: ὅρασιν, ἀκοήν, ὄσφρησιν, γεῦσιν καὶ ἁφήν. Marcovich renders it: ὅρασιν, [ἀκοήν,] γεῦσιν, ὄσφρησιν καὶ ἁφήν. But his emended text contravenes the order of the senses presented at 6.15.2–6.16.4: sight, hearing, smell, taste, and touch. This latter order follows exactly that of Plato *Timaeus* 61–67 (intentionally reversed); Chrysippus, frags. 827, 836 (SVF 2:226–227); Aetius *Placita* 4.9.10 (= Stobaeus *Eclogae* 1.50.27 [Wachsmuth and Hense 1894:476]); and others (see e.g. Lampros 1895:2.17, no. 4212.72). The same order is preserved in the independent

Thus deutero-Simon synthesizes into his theory of sense perception two very different biblical numbers: four (the number of rivers in paradise) and five (the number of books in the Pentateuch). This uneven treatment of the senses may be inspired by Plato's *Timaeus*, where taste, smell, hearing, and sight are treated as a quartet, separate from touch, and in the same order, albeit in reverse. If so, then Deutero-Simon, like Monoïmus, wanted to engage contemporary philosophers and the cultural elite.

Deutero-Simon also attempts, as does Monoïmus, to understand the recurrent patterns of five and ten in the Pentateuch in anthropological terms. But whereas Monoïmus connects the tens in Scripture to Aristotle's ten categories and the shape of the iota, deutero-Simon connects the five senses to the four rivers and the five books of Moses. In other respects Monoïmus and deutero-Simon are similar. Monoïmus calls the μία μονάς "many-faced and ten-thousand–eyed and ten-thousand–named" (πολυπρόσωπος καὶ μυριόμματος καὶ μυριώνυμος; 8.12.7); the *Apophasis Megale* seemingly says the Son is "many-named, ten-thousand–eyed, incomprehensible" (πολυώνυμος μυριόμματος ἀκατάληπτος; 5.9.4).[54] Both deutero-Simon and Monoïmus attempt to relate the seven days of creation to seven powers latent in a transcendent realm (termed 'Infinite Power' in deutero-Simon and 'Human' in Monoïmus). These powers emerge first as a set of six, resembling the core of the Pleroma taught by Hippolytus' Valentinians, and then the seventh follows. Monoïmus' system does not teach the syzygies found in deutero-Simon's, but their shared arrangement of the seven powers in groups of six and one is striking.

Deutero-Simon's syzygies resemble those of the Valentinians.[55] But unlike the syzygies in the models of Valentinianism reported by Irenaeus, all the

---

but parallel accounts at Irenaeus *Against Heresies* 1.18.1 and Hippolytus *Refutation of All Heresies* 5.9.16–18, where the first four are assigned, in that order, to the four rivers. Because these parallel texts omit touch, so as to match the symbolism of four, it is likely that the *Paraphrase of the "Apophasis Megale"* does so as well at 6.15.1.3. My intuition is justified by an observation made much earlier by Salles-Dabadie 1969:28n17, the only modern editor to seclude καὶ ἁφήν: in deutero-Simon's system the fifth book of the Torah is called *Deuteronomy* to supply to an already formed child—presumably after birth—touch, the capstone of the senses (Hippolytus *Refutation of All Heresies* 6.16.3): Δευτερονόμιον δὲ τὸ πέμπτον βιβλίον, ὅπερ, φησίν, ἐστὶ πρὸς τὴν ἁφὴν τοῦ πεπλασμένου παιδίου γεγραμμένον. ὥσπερ γὰρ ἡ ἁφὴ τὰ ὑπὸ τῶν ἄλλων αἰσθήσεων ὁραθέντα θιγοῦσα ἀνακεφαλαιοῦται καὶ βεβαιοῖ, σκληρὸν ἢ γλίσχρον, ἢ θερμὸν ἢ ψυχρὸν δοκιμάσασα, οὕτως τὸ πέμπτον βιβλίον τοῦ νόμου ἀνακεφαλαίωσίς ἐστι τῶν πρὸ αὐτοῦ γραφέντων τεσσάρων. Thus, καὶ ἁφήν at 6.15.1.3 was inadvertently inserted at the list's end, where many such scribal intrusions occur, and the text should read ὅρασιν, ἀκοήν, ὄσφρησιν, ‹καὶ γεῦσιν› [καὶ ἁφήν]. For departures from the canonical order of the senses, see p. 129n20 below.

54   "Seemingly" because the latter phrase occurs in Hippolytus' discussion of the "Phrygians," but just before an explicit reference to the *Apophasis Megale*. Hippolytus may have introduced the phrase in anticipation of the section to come in Book 6.

55   See Marcovich 1986:217n7 for an extensive list of comparative references.

syzygies in the *Paraphrase* emanate from the Infinite Power, not one from another. There is no hint in the *Paraphrase* of a doctrine of ogdoads, decads, dodecads, or triacontads. Instead, the *Paraphrase* emphasizes the number seven, a number not given much significance in Valentinianism, except as a symbol of the rather lowly Demiurge. Further, deutero-Simon's naming scheme differs from the Valentinians'. The three male powers describe human faculties (Mind, Voice, Reason), and the names of the three female powers are the product or result of their male counterparts (Thought, Name, Consideration). These are the pairs because a Mind generates Thought; a Voice utters a Name; and Reason gives rise to Consideration. There is no direct connection to the various Valentinian naming schemes, despite the similar logic of using them to expound the actions or properties latent within the Monad.

Nothing suggests that the *Paraphrase*, Monoïmus, or the Valentinians depended on one another for their number symbolism. Nor did they need to. Numerical theology was a broad trend in late second- and early third-century Christianity, not restricted to any one movement. It included Christians who, inspired by neo-Pythagoreanism and Platonism, developed an arithmetically shaped mythology to provoke and persuade both churchmen and the cultural elite of a transcendent vision. Numerous theologies of arithmetic circulated, the Valentinian ones being the best preserved. Some models tried to preserve in their metaphysical narrative a simple or pure philosophical idea. Some tried to synthesize competing models. Others combined protologies eclectically, with no regard for reconciling contradictions. Every system presents a story of how the highest entity (or entities, if a Dyad) generated mathematically patterned multiplicity. In most of these systems numbers were repeatedly used to interpret the Bible, the natural world, and even social conventions such as grammar and names. In this they were working with techniques of arithmological interpretation familiar in the Greco-Roman world. But familiarity did not entail plausibility. Orthodox Christians argued for another way.

6

The Orthodox Limits of the
Theology of Arithmetic

Irenaeus of Lyons

MOST SCHOLARLY RECONSTRUCTIONS OF VALENTINIANISM attempt to excise Irenaeus' interpretation, so as to achieve as uncontaminated an account of the movement as possible. The ultimate goal, a bias-free view of the Valentinians, is misguided. For the unintended consequence—or so I argue in this chapter—of excluding Irenaeus and his writings is to leave the picture of their number symbolism incomplete. Irenaeus' critique of Valentinian theological arithmetic reveals the rules and restraints held in common by many Christians. And his vision for the appropriate use of number symbolism illuminates the general early Christian impulse to number symbolism that the Valentinians shared. Irenaeus' ideals, and his success and failure in holding to them, provide a glimpse into tensions in early Christian thought, both orthodox and heretical. So to understand the Valentinians, we must study Irenaeus, too, on his own terms.

Born in the early second century and growing up in Asia Minor, Irenaeus listened to the teaching of Polycarp of Smyrna (d. 156), the famous martyred bishop who, in turn, had reportedly learned at the feet of the apostle John.[1] Irenaeus made the most of his status as a spiritual grandson of the apostles. His early interaction with Polycarp and other local elders, and his immersion in the Christian community of Asia Minor, shaped his career as a priest and bishop in Lyons, where he remained from the 170s until the end of his life, probably around 200. He was popular with his Gallic Christian flock, who suffered persecution during his tenure. He gained Church-wide renown from his attempts to reconcile the Roman church with other Christians who used a different calculation for Easter. Little else is known of his life, aside from what Eusebius reports.[2]

---

[1]   Irenaeus *Against Heresies* 3.3.4.
[2]   See DECL, s.v. "Irenaeus"; Eusebius *Church History* 5.4, 7–8, 20, 26.

Only two of Irenaeus' many known works are preserved complete: *Against Heresies* and *Demonstration of the Apostolic Preaching*. Of his lost treatises, one, titled *On the Ogdoad*, was directed against a certain Florinus, who broke away from the church in Rome and later joined Valentinian circles. This attests to Irenaeus' unflagging interest in and opposition to the Valentinians, and perhaps his obsession with their theology of arithmetic.[3] But since we no longer have that treatise, we must rely upon *Against Heresies* to reconstruct his views on the Ogdoad and on Valentinian number symbolism in general.

Setting aside the rhetorical flourishes expected in ancient theological polemic—Irenaeus' often wry sarcasm and labeling (e.g. his accusation that Valentinianism derives from Pythagoreanism and paganism)—he advances four general theses against the Valentinian use of numbers. First, the aeons in their Pleroma are inconsistently numbered; second, their doctrine makes claims that depend upon the changing, culture-bound customs of language and numeration; third, their number symbolism does not correspond to the structures of the created, natural world; and fourth, their method of interpreting the rule of faith, Scripture in particular, is faulty. While pursuing these four lines of attack, Irenaeus promotes some basic principles about theology and exegesis, both to discredit his opponents and to articulate core principles of the faith held by the churches around the world.

Irenaeus' sarcastic asides are of little substance. Yet his colorful insults provide context for his substantive points. He mocks the Valentinian Tetrad by inventing his own out of an emptiness, a gourd, a cucumber, and a melon (*Against Heresies* 1.11.4). This emanation of fruits, Irenaeus argues, is just as plausible as that of their aeons, and equally arbitrary. While reporting Marcus' system, Irenaeus accuses him of trying to preach something more mystical than the other systems, of achieving new spiritual heights by breaking everything down into numbers. He also associates his opponents with pagan teachings and habits. A Pythagorean slur occurs in Book 2, where Irenaeus claims that the Valentinian tendency to translate everything into numbers comes from the Pythagoreans (2.14.6, upon which the rest of this paragraph is based). His logic: they were the first to make numbers the origin of everything, the first to make even and odd the foundation of numbers, and the first to make odd and even the basis of sensible and intelligible things. Even numbers are the basis for underlying substance (in Aristotelian terms, a 'primary substance'), whereas odd numbers are the basis for intellection and essence.[4] The difference between even and odd resembles the parts of a statue, which has both substance (equivalent to

---

[3] Eusebius preserves the closing words of the treatise: *Church History* 5.20.2.
[4] The preserved text is muddled here. My paraphrase follows the conjecture of Rousseau et al. 1965–1982:1.260.

even numbers) and form (odd numbers). This is the sort of model, he says, that the Valentinians apply to beings outside the Pleroma. They (the Pythagoreans or the Valentinians—the text is vague) claim that by knowing "what was first assumed" (*quod primum adsumptum est*), a person seeks out the beginnings of intellection and in exhaustion races to that which is one and indivisible. This One—the ἕν—is the principle of everything and the basis of all generation. From it come the Dyad, Tetrad, and Pentad—terms that the Valentinians use to describe the Pleroma and Depth. This Pythagorean number symbolism undergirds their doctrine of the syzygies. Marcus, boasting about the great novelty of his invention, speaks about the τετρακτύς of Pythagoras as if it were the origin and mother of everything.

Unlike Hippolytus, who establishes simplistic one-to-one correspondences between various philosophers and heresiarchs,[5] Irenaeus links Valentinianism to all the various strains of Hellenistic thought, claiming their system to be a pastiche of Homer, Hesiod, Democritus, Epicurus, Anaxagoras, Empedocles, Plato, Aristotle, and the Cynics, as well as the Pythagoreans. To cap the slander, he claims that his opponents draw inspiration from the pagan pantheon of twelve gods, and make them images of the Dodecad (2.14.9).[6]

Such ridicule and insults are mere flourishes to Irenaeus' more substantive, theological arguments, which he presents directly and forcefully. Using counterexamples drawn from Scripture, history, and the natural world, he insists that the Valentinians—exemplified by the extended system of the Triacontad—have not correctly counted the number of aeons in the Pleroma (the first of his four main arguments). To make this point, he defends two claims. First, by their own reckoning, their Pleroma has fewer than thirty aeons. Second, the Pleroma ought to have more than thirty aeons. In each case, the Valentinian school is shown to be incapable of responsibly handling the numbers it so esteems in its theology.

In the first of these arguments, that there are really fewer than thirty aeons, Irenaeus first focuses on the role of the Forefather (2.12.1). If he is the source of the various projections, he ought not to be counted with them, since we should not group one who emits, who is unbegotten, who is neither circumscribed nor given form, with one who is emitted, begotten, circumscribed, or formed. Likewise, the Forefather should not be grouped with Wisdom, since this is to group together an errant aeon with an inerrant one. To classify them together is to suggest they share the same nature, an impossibility given Wisdom's fall

---

[5]   See p. 55 and n81 above.

[6]   This may be Irenaeus' invention, or it may reflect an analogy the Valentinians themselves made between the Dodecad and the gods or the zodiac. The latter possibility should not be discounted, given the Valentinian penchant for Pythagorean symbolism.

from the Pleroma. So if we were to exclude the Forefather and Wisdom from the Pleroma, as seems appropriate, we would end up with only twenty-eight aeons.[7]

And what about Silence (Thought), Depth's consort (2.12.2)? Can any being be separated from its own silence and thought? Indeed, does not the very notion of conjugal unity forbid any idea of separation? If so, then Thought is in every way similar to Depth; they share a single existence. This applies as well to the other conjugal pairs. Mind and Truth, who always indwell one another, cannot be separated, just as water and moisture, fire and heat, and stone and hardness cannot be separated. Likewise, Word and Life, Human and Church, and all the other pairs of aeons cannot be disentangled. After all, the feminine aeon must necessarily be equal to the masculine, "since the former resembles the latter's disposition."[8] 'Disposition,' a Valentinian term, implies metaphysical unity. So we should count only syzygies, not aeons. Irenaeus anticipates their response, that the syzygies are in fact divided, so that individual aeons can be enumerated apart from their mates (2.12.4). But, Irenaeus charges, this renders absurd their other claim that the syzygies are unities and that the male and female are one. If the aeons of a given syzygy are separate, then the female gives birth to offspring apart from her mate. If so, then she is like a hen who hatches eggs without the help of a rooster. Irenaeus' argument here boils down to whether the syzygies are real or only symbolic, and whether they are to be counted singly or doubly. The Valentinians' inability to specify whether the syzygies are true unions or only token ones, he says, makes it impossible to tell whether they have counted accurately.

Irenaeus also argues that the Valentinians' system results in more than thirty aeons in their Pleroma (2.12.7, upon which this paragraph is based). They say that four other entities are projected—Limit, Christ, Holy Spirit, and Savior (see Figure 3, p. 38 above)—but they do not include them in the canonical thirty of the Pleroma. Why not? Irenaeus asks. Are they so weak as to be unworthy of the designation? Are they superior to the other aeons? It would be absurd to suggest that they are weaker, since they were projected to stabilize and correct the Pleroma. But it would be equally absurd to suggest that they are better than the primal Tetrad. So, if they are neither weaker than the weakest aeons in the Pleroma, nor better than the best, then either they should be numbered with

---

7   The logic is similar to that employed by Hippolytus' Valentinians, who derive the same number by excluding the Forefather from the class of aeons.

8   *cum sit velut adfectio eius.* Here, *adfectio* probably represents διάθεσις. See SC 293, index, s.vv. 'διάθεσις' and '*adfectio.*' To describe the female aeon as the διάθεσις of the male is characteristic of the "more knowledgeable" Ptolemaeans discussed at *Against Heresies* 1.12.1. See pp. 35–37 above.

the Pleroma, or the honor associated with such a name (πλήρωμα means "full-ness") should be removed from the other aeons since, obviously, the Pleroma does not include the fullness of the aeons.

Irenaeus' second line of attack calls Marcus to account for his misuse of numeration conventions. He ridicules the notion that the Word the Father uttered consisted of thirty letters and four syllables (1.15.5). If this were so, then the Father, whose image the Word bears, should also consist of thirty letters and four syllables. Indeed, is this really the final arrangement? Marcus, says Irenaeus, bottles up the Creator in various numbers and patterns: at one time thirty, at another twenty-four, at another merely six. Even the technique he uses to calculate alphabetic numbers is inconsistent, since at one time he computes a name's psephic value, at another time the number of letters in the word (2.24.1–2). 'Jesus,' Irenaeus points out, is not a Greek name, yet Marcus never-theless makes its Greek transliteration the center of his theology. Sometimes he calls it the ἐπίσημον because it has six letters, and sometimes the full-ness of the Ogdoad, since the psephic value of Ἰησοῦς is 888. But even if this were true, he does not do this with the Lord's other names and titles, such as Σωτήρ. And no wonder, since neither its psephic value, 1,408, nor the number of letters in it, five, is related to the numbers or patterns in the Pleroma. The psephic value of 'Christ,' Χρειστός, is 1,485, but this has no arithmetical connec-tion with the Pleroma that Christ allegedly stabilizes and corrects. The same applies to Πατήρ, Βύθος, Μονογενής, and other aeons. These inconsistencies— applying Greek linguistic conventions to Hebrew names and not applying the same method to the more important Greek names—proves that their system is false.

Irenaeus argues that the system does not square with the history of the alphabet. He says that the Greeks agree: only recently—recently, that is, relative to the creation of the world—Cadmus introduced the first sixteen letters (1.15.4). Some time after, other Greeks invented the aspirates (θ, φ, χ) and the double letters (ζ, ξ, ψ). Palamedes provided the long vowels (η, ω), the final step in the alphabet's evolution. Thus, Irenaeus argues, how could Marcus' Truth exist before the rise of the Greek alphabet, seeing that her body had to have postdated Cadmus? Indeed, Truth postdates "even yourself [Marcus], for you alone have dragged down your so-called 'Truth' as an idol."[9] Although Irenaeus' version of the history of the alphabet is overly confident—the specifics about who invented the Greek alphabet, and when, vary from one ancient author to the next—his overall point that the Greek alphabet had a progressive origin is correct, both by his

---

[9] μεταγενέστερον δὲ καὶ σαυτοῦ· σὺ γὰρ μόνος <εἰς> εἴδωλον κατήγαγες τὴν ὑπὸ σοῦ λεγομένην Ἀλήθειαν.

generation's understanding of the history of the alphabet and by that of modern scholars.[10]

Irenaeus charges Marcus with failing to abide by the conventions appropriate to a given language (2.24.2). 'Jesus,' a Hebrew name, properly consists in its source language of two and a half letters, a claim Irenaeus justifies by appealing to Jewish experts, who take each of the letters in ישׁו (instead of the biblical ישׁוע) as an acronym for "Lord, heaven, earth." (Here the *yod* seems to be counted as the half letter.) Thus, just as Σωτήρ exposes the inconsistency of their system, so too does Jesus' Hebrew name. Its mere two and a half letters show that 'Jesus' cannot be considered the ἐπίσημον. The interpretation of 'Jesus' as 888 too cannot be sustained in the context of the name's origin. Hebrew letters do not match Greek letters, and because the former are older and more stable than the latter, any calculation of names should be based upon the older. And besides, the very structure of the Hebrew alphabet precludes any kind of psephy.[11]

According to the Valentinians, the Demiurge fashioned the natural world as an image of the unseen Pleroma.[12] This proposition leads to Irenaeus' third argument against the Valentinians, that the numbers in their system do not correspond to what we know of the natural world. Irenaeus pursues this attack numerous times, in every case criticizing as illogical the notion that an errant being, the Demiurge, could create the world (itself errant) as a true reflection of an inerrant Pleroma (2.7.1–6). He anticipates the response that the natural world is the image of the Pleroma, not in figure or form, but in number and rank (2.7.7). Irenaeus answers that not even this is true, since the Valentinians tend to tinker with their numbers and their aeons so as to make them fit creation. But even then, granted that they have managed to make some associations with the natural world, how can they claim on this basis that a mere thirty aeons

---

[10] See Pliny the Elder *Natural History* 7.192, Tacitus *Annales* 11.14, and many others cited at Förster 1999:238–242 and Teodorsson 1989–1996:3.318. Irenaeus' report resembles that of Plutarch's *Table Talk* 9.3 (738–739). For modern approaches to the history of the Greek alphabet, see OCD s.v. "Alphabet, Greek."

[11] This is my interpretation of *Against Heresies* 2.24.2.40–46, a convoluted passage poorly preserved in the Latin: *Ipsae enim antiquae et primae Hebraeorum litterae et sacerdotales nuncupatae x quidem sunt numero: scribuntur autem quaeque per xv, nouissima littera copulate primae. Et ideo quaedam quidem secundum subsequentiam scribunt, sicuti et nos, quaedam autem retrorsum a dextera parte in sinistram retorquentes litteras.* As we have it, the text states that Hebrew has ten letters, each written "through fifteen, the more recent letters joined with the first." The text also seems to appeal to the direction of writing, from right to left. The earlier part of 2.24.2 deals with two numerical practices regarding names: the number of letters in a word, and its psephic value. Irenaeus dispenses with the first notion by highlighting the peculiar way (he says) letters are counted in Hebrew. This muddled and unparalleled explanation of the structure of the Hebrew alphabet must then refute Marcus' claims by showing that his dependence upon Greek psephy has no logic in the Hebrew alphabet.

[12] Treated passim in Irenaeus *Against Heresies* 1. See pp. 42–44 and 52 above.

are the antitype? The enormous numerical complexity of the creation cannot be explained merely by a group of thirty entities. The world is no image of the Pleroma.

Later in Book 2, Irenaeus extends this argument (2.15.1). The Valentinians, he says, claim that the thirty aeons were not made for creation but vice versa. That is, the creation is the image of the thirty aeons, not the other way around. According to their reasoning, the month has thirty days because of the thirty aeons. So too the day has twelve hours, and the year twelve months, because of the Dodecad. The reasoning is arbitrary and incomplete, Irenaeus says, since it does not explain why Human and Church had to project twelve, no more and no less. It also does not explain why an Ogdoad, and not, say, a Pentad, Trinity, or Heptad, is the core of the Pleroma. If the year is an image of the Dodecad, and the month of the Pleroma, then of what important natural occurrence of the number eight is the Ogdoad an image?

Irenaeus accuses the Valentinians of using analogies that invert the order of nature (2.24.5). They appeal to the divisions of the year and the day as symbols of the Pleroma. The number of months and the number of hours of the day point to the Dodecad; the days in the month, to the Triacontad. The scheme is inconsistent. Each aeon is supposed to be one-thirtieth of the Pleroma, but a month is one-twelfth of a year. If the divisions of time really reflected those of the Pleroma, wouldn't it have been more appropriate for the year to be divided into thirty months and each month divided into twelve days? The Savior must have been an idiot, Irenaeus chides them, to have made the month an image of the Pleroma and the year, the more important division of time, that of the Dodecad, a less important subset of the Pleroma. And their analogy does not account for the realities of the calendar. Not every month has precisely thirty days, just as not all days have twelve hours, depending upon the season of the year. Thus neither can the day be a true image of the Dodecad, nor can the month of the Pleroma. And why do they group the Pleroma into Ogdoad, Decad, and Dodecad and no other arrangement? (2.15.2) Why three divisions, and not four, five, six, or some other number more commonly found in creation? After all, the year is divided into four seasons (2.24.5). Were the year truly an image of the Pleroma, the aeons would fall into four major tiers, not three.

Whenever they are cross-examined about the Pleroma, Irenaeus says, they retreat to explanations about human dispositions and to discourses about creation (2.15.3). But this is to focus on secondary rather than primary matters, since the issue at hand is not harmony in creation or human dispositions, but the Pleroma, of which creation is an image. If the Pleroma is trisected into Ogdoad, Decad, and Dodecad, then they must admit that the Father arranged the Pleroma in vain and without providence, since its organization did not correctly anticipate the structures of the natural world. If so, then the Father acted irrationally and, like the Demiurge, made a deformity. Given their analogy between the Pleroma and creation, they must admit that the Forefather is just as inept as they say the creator of the natural world is. But if they want to avoid that conclusion and instead maintain the providence of the Forefather, then they must say that the Pleroma was projected so as to provide a template for creation. But in this case, although the harmony of the cosmos is preserved, the Pleroma exists not for itself but for its image. The Pleroma is then inferior to creation, as if it were a clay model, made only to facilitate the construction of a gold, silver, or brass statue. In sum, if the Pleroma is a template for creation, then the Father's creation is inferior to what the Demiurge made.

Irenaeus accuses the Valentinians of fostering an infinite regression that wreaks havoc with their arithmetical theology. If they don't agree that the Pleroma was made for creation, then they must say that it was made for a higher reason or cause (2.16.1). But this is to postulate that Depth used a higher pattern to shape and arrange the Pleroma. A vicious regression begins, since you must ask how and why this super-Pleroma was made. The same line of questioning can lead to super-super-Pleromas, and so on. Irenaeus argues that you must accept that one God, who made the world, took the pattern for creation from his own power and his own self. If you deviate the least bit from monotheism, you end up always asking and seeking out how and from what source the creator patterned his creation, upon what he styled the number of projections, and from where he derived the substance that was used. If they try to argue that Depth perfected the design of the Pleroma from himself, then it should be allowed that perhaps the Demiurge also patterned the world, not from the Pleroma, but from himself. That is, maybe the Demiurge is not so evil after all. But if they insist that creation is an image of the Pleroma, then what is to prevent the Pleroma from also being an image, and so descend into an endless regression of images of images?

Throughout his argument, Irenaeus focuses on Valentinian number symbolism. He pursues the numerical patterns and structures central to their theology to test their claims that there are innate, direct connections between creation and the Pleroma. That there are differences between natural and divine patterns raises difficult questions. Why those numbers in particular? Where did they come from? Are they intrinsic to the highest reality, or merely accidental parts? Irenaeus argues that the heretics make number external to the deity, which is why they always try to trump each others' protologies, since deity is subservient to unbounded, arbitrary arithmetical patterns. And the origin of those patterns is troublesome. Basilides (2.16.2) says that Ineffable based the pattern for the emanation of the heavens upon his dispensation. But where did that dispensation come from? Ineffable had to have created the dispensation either from himself, or from a yet higher power. If the former, then one might as well abandon the entire theological complex of aeons in favor of simply the one God, who created the one world from a numerical pattern of his own design (2.16.3). If the latter, then the endless regression resumes.

Irenaeus mocks his opponents' increasingly complex systems as symptomatic of their spiritual arrogance (2.16.4). Just as the Valentinians accused less spiritual Christians such as Irenaeus of remaining in the Hebdomad, so the Basilideans could accuse the Valentinians of remaining at the level of the Triacontad, and not ascending to the 45 ogdoads, then the 365 heavens. The higher the number, the better, no? Why not invent a system of heavens or aeons numbering 4,380, the number of daytime hours in a year? And then build upon the nighttime hours an even greater number? This endless one-upmanship means that the Valentinians and Basilideans will always be unable to rise to the highest conception of the number of heavens or aeons. The creation of such tiers above our world is an invitation to descend into endless levels of worlds (2.35.1).

Irenaeus returns to this theme in Book 4 of *Against Heresies*, where he advises that anyone who seeks a Father beyond the one Father of the Scriptures will need to find a third, a fourth, and so on (4.9.3). Such a person will never rest in the one God, but will drown in a Depth without Limits, until repentance brings him back to the place from which he was cast out, the one God. In his admonition, Irenaeus symbolizes God by the garden of Eden, the place of single simplicity. To develop other gods or aeons is to begin a journey into a bottomless pit of multiplicity, a transience ended only by returning to the garden of God's unity.

In his fourth line of argumentation, Irenaeus deals with the numerous specimens of Valentinian numerical exegesis presented in Book 1, attacking their hermeneutical principles as arbitrary and inconsistent. According to the Valentinians, the prologue of the Gospel of John justifies the names and sequence of the aeons of the Ogdoad (1.9.1).[13] But, Irenaeus answers, John introduces the terms in an order quite different from theirs. If the structure of their Pleroma is so important, surely John would have preserved this sequence, documented their conjugal unions, and mentioned every aeon by name (Church, Human's consort, is never explicitly mentioned in John 1).

To the notion that Judas represents Wisdom, the fallen twelfth aeon, Irenaeus responds (2.20.2):[14] Judas was indeed expelled, but never reinstated, as Wisdom allegedly was. Since Matthias took Judas' place (Acts 1.20), their

---

[13] See p. 45 above.
[14] See p. 46 above.

myth ought to follow suit, and have another aeon projected to replace Wisdom. Furthermore, they say that Wisdom suffered, but then they themselves admit that Jesus, not Judas, suffered. How can an unsuffering traitor be the image of a suffering aeon? After mentioning other dissimilarities between Judas and Wisdom, Irenaeus takes the Valentinians to task for their counting mistakes (2.20.4). True, Judas is the twelfth, but they teach that Wisdom was the thirtieth. How could a twelfth-ranked Judas be a type of the thirtieth-ranked aeon? Even if you accept the Judas-Wisdom connection, other problems in enumeration occur (2.20.5). They say that Judas' death represents the Inclination of Wisdom, an entity that, in their myth, returns to the Pleroma. But this cannot resemble Judas, who was never reinstated with the apostles. In the course of their explanation they try to use two biblical figures, Judas and Matthias, to stand in for three aeons, Passion, Inclination, and Wisdom. But two cannot equal three.

Moreover, if the Valentinians want the twelve apostles to represent Human and Church's projected twelve aeons, to be consistent they should produce ten more apostles to represent the other Decad of aeons, emitted by Word and Life (2.21.1). How could the Savior provide a type for the youngest (and therefore least significant) aeons, but overlook the elder ten? The same applies to the Ogdoad, which ought to have been numerically signified by the election of eight apostles. Indeed, their system can make nothing of the seventy other apostles the Lord sent after he commissioned the twelve, since seventy prefigures neither an Ogdoad nor a group of thirty. If their reasoning is correct, that the election of the twelve apostles signifies the twelve aeons, then they must hold that the seventy apostles were chosen because of seventy aeons. If this is the case, then the two groups of apostles, twelve and seventy in number, symbolize eighty-two aeons, far beyond the canonical thirty.

The Valentinians claim that the woman with the twelve-year flow of blood symbolizes the restoration of Wisdom (1.3.3).[15] Irenaeus argues that the story is inconsistent with their system, which holds that eleven of the twelve aeons in the Dodecad were unaffected by suffering, and that only the twelfth suffered (2.23.1). But the woman who was healed experienced the opposite. She suffered for eleven years and was healed in the twelfth. Irenaeus concedes that a type or image differs from the truth it represents according to the material and underlying substance, but the type must nevertheless preserve the form and outline of the truth. A type should make evident by its presence that which is not present.[16] Furthermore, the Valentinians do not apply this exegetical prin-

---

[15] See p. 46 above.

[16] Recall Irenaeus' analogy of the relationship between a clay model and the gold statue upon which it is made (2.15.2, discussed at p. 110 above). The two objects have different material causes, but share an identical form.

ciple consistently. What do they do with the woman who suffered for eighteen years (2.32.2, citing Luke 13.11)? If one woman is a type of the aeons, then the other should be, too. The same goes for the man who was healed after being sick for thirty-eight years (John 5.5). Since these two examples have no bearing on their system, then neither should the first.

To the Valentinian appeal to the numbers in the Mosaic Law as symbols of the Pleroma, Irenaeus offers numerous counterexamples (2.24.3). Their system, he asserts, has nothing to say about the dimensions of vessels in the holy of holies: the ark of the covenant (2½ by 1½ by 1½ cubits), the mercy seat (2½ by 1½ cubits), and the table of the showbread (2 by 1 by 1½ cubits). The seven-branched candelabra, by their scheme, ought to have been made with eight lights, to typify the primal Ogdoad (Exodus 25.31–37). They appeal to the 10 curtains (Exodus 26.1) as a type of the 10 aeons, but they neglect the skin coverings, 11 in total (Exodus 26.7), and the length of the curtains, 28 cubits (Exodus 26.2). Although they say the 10-cubit length of the columns typifies the Decad of aeons, they cannot explain their 1½-cubit width (Exodus 26.16). Furthermore, they cannot account for the oil (500 shekels of myrrh), the cassia (also 500), the cinnamon, and the calamus (both 250). These four items plus the oil make five ingredients, a number that does not fit their scheme.

Irenaeus argues that the Valentinians' application of finger calculation and psephy is inconsistent (2.24.6).[17] To say that the Savior came to gather the hundredth lost sheep and to transfer to the right hand the 99 that are on the left does not mesh with their own analogy. According to them, anything on the left belongs to corruption, in which case it is not the hundredth but the 99 who are lost, since they are the ones who exist on the left hand. Furthermore, they think that anything that does not carry at least the number 100 belongs to the left side, of corruption. So since the psephic value of ἀγάπη 'love' is 93, it too must reside on the left side, as must ἀλήθεια 'truth', which is 64, and anything else whose psephic value is less than 100.[18]

To show that Valentinian exegesis of the numbers in Scripture is capricious, Irenaeus mocks it by glorifying the wonders of five, a number that fits nowhere in the structures of Valentinian theology (2.24.4). Five recurs throughout Scripture. Σωτήρ, Πατήρ, and ἀγάπη all have five letters. The Lord, blessing the five loaves, feeds the five thousand. There are five wise virgins and five foolish.

---

[17]   See pp. 30 and 80 above.

[18]   Marcus' differentiation between numbers greater than and less than 100 resembles techniques in second-century dream interpretation, which treated 100 as an auspicious number and evaluated the names of things occurring in dreams accordingly. Artemidorus *Dream Book* 2.70, 3.34, says that numbers should be reduced to 100 to be suitable for interpretation, and that the number 100 is especially auspicious. But Artemidorus does not consider numbers less than 100 to be inherently unlucky.

There are five men with the Lord at the Transfiguration. The Lord is the fifth of those who entered the house of the ruler whose daughter was ill (Luke 8.51). The rich man in the infernal regions says he has five brothers (Luke 16.19–31). The pool at Bethesda has five porticoes (John 5.2). The cross consists of five parts: the four arms and the center. Every hand has five fingers and there are five senses and five internal organs: heart, liver, lungs, spleen, and kidneys. There are five divisions in the body and five phases in human life.[19] Moses gave the Law in five books and each tablet contained five laws.[20] Five priests were elected in the desert—Aaron, Nadab, Abiud, Eleazar, and Ithamar—and the ephod and breastplate were made of five materials—gold, hyacinth, porphyry, scarlet, and fine linen. Joshua surrounded five kings of the Amorites. The list could go on, Irenaeus notes, but this vast array of fives is no reason for claiming a divine group of five aeons.

Irenaeus turns the Valentinians' technique against them. He notes sarcastically that according to the Scriptures, an exorcised spirit of ignorance finds those who were formerly possessed "striving not after God but after cosmic inquiries, and brings along seven other spirits more wicked than himself" (1.16.3, citing Matthew 12.43). One spirit plus seven others makes eight. By the Valentinians' own logic, a demon has left them but returned and found them ready to be inhabited, and so has taken along "seven other spirits, thus constituting their Ogdoad of spirits of wickedness." Their arrogance is so great that, in enumerating seven heavens, they presume to have surpassed the apostle Paul, who ascended only to the third heaven, four short of the highest level (2.30.7, citing 2 Corinthians 12.2).

Number Symbolism in Irenaeus' Theology and Exegesis

All four of these lines of argumentation boil down to charges of incompleteness, inconsistency, and arbitrariness. But someone must have pressed Irenaeus to offer his own account of how numbers should be used responsibly in theology and biblical interpretation. We get a hint of this when the text asks: Are the

---

[19] Irenaeus' list: infancy, childhood, youth, adulthood, and old age, is shorter than Solon's (fragment 19 Diehl 1971) and Hippocrates' (*On Hebdomads* 5) more famous model, of seven phases in human life. But cf. Theon of Smyrna *Mathematics Useful for Reading Plato* 98.13–14, which divides human life into four, a τετρακτύς.

[20] Note how Irenaeus' proofs move from the New Testament to the natural world to the Old Testament. This mirrors the order and sequence of the Valentinian exegesis Irenaeus presents in Book 1, where the New Testament proof texts for the Pleroma are at chaps. 1–3 and 8, natural world proofs are at chap. 17, and Old Testament proof texts at chap. 18. This is evidence both for the sequence of Irenaeus' source and for the basic integrity of the authorship and sequence of *Against Heresies* 1.1–22.2, which may constitute Irenaeus' first draft of Book 1. See Kalvesmaki 2007b.

placement of names, the election of apostles, and the acts of the Lord and his deeds recorded for no reason whatsoever? Irenaeus replies:

> Not at all. Rather, everything God does—whether ancient or anything accomplished by his Word in recent times—is harmonized and well ordered with abundant wisdom and precision. And these things should be yoked not with the number thirty but with the underlying narrative of truth. And they should not undertake an investigation about God on the basis of numbers, syllables, and written letters. (For this is unsound because of their multifaceted and variegated nature, and because any narrative—even one that someone cooks up today—can gather out of the same [numbers, syllables, and letters] proof texts contrary to the truth, in that they can be manipulated to many ends.) Rather, they should fit to the underlying narrative of truth the numbers themselves and the things that have been done. For the rule does not come from numbers, but numbers from the rule (*non enim regula ex numeris, sed numeri ex regula*). Neither does God come from creation, but that which is made, from God. For everything is from one and the same God.

(Irenaeus *Against Heresies* 2.25.1)[21]

[21] "Non quidem, sed cum magna sapientia et diligentia ad liquidum apta et ornata omnia a Deo facta sunt, et antiqua et quaecumque in novissimis temporibus Verbum eius operatum est. Et debent ea, non numero xxx, sed subiacenti copulare argumento sive rationi, neque de Deo inquisitionem ex numeris et syllabis et litteris accipere—infirmum est enim hoc propter multifarium et varium eorum, et quod possit omne argumentum hodieque commentatum ab aliquo contraria veritati ex ipsis sumere testimonia, eo quod in multa transferri possint—sed ipsos numeros et ea quae facta sunt aptare debent subiacenti veritatis argumento. Non enim regula ex numeris, sed numeri ex regula, neque Deus ex factis, sed ea quae facta sunt ex Deo: omnia enim ex uno et eodem Deo." Rousseau et al.'s reconstruction of the Greek (1965–1982), with my own conjectures in angle brackets: Οὐ μὴν ἀλλὰ μετὰ μεγάλης σοφίας καὶ ἀκριβείας εὔρυθμα καὶ ἐγκατάσκευα πάντα ὑπό τοῦ Θεοῦ ἐγένετο, τά τε ἀρχαῖα καὶ τὰ ὅσα ἐν ἐσχάτοις καιροῖς ὁ Λόγος αὐτοῦ ἔπραξεν· ὀφείλουσι δὲ αὐτὰ μὴ τῷ ἀριθμῷ τῶν τριάκοντα <τῷ τριακοντάδι?>, ἀλλὰ τῇ ὑποκειμένῃ συνάπτειν <συνεικειοῦν?> ὑποθέσει τῆς ἀληθείας, μηδὲ περὶ τοῦ θεοῦ ζήτησιν ἐξ ἀριθμῶν καὶ συλλαβῶν καὶ γραμμάτων ἀναδέχεσθαι—ἀσθενὲς γὰρ τοῦτο διὰ τὸ πολυμερὲς καὶ πολυποίκιλον αὐτῶν καὶ τὸ δύνασθαι πᾶσαν ὑπόθεσιν καὶ σήμερον παρεπινοουμένην ὑπό τινος ἀναλήθεις ἐξ αὐτῶν λαμβάνειν μαρτυρίας ἅτε εἰς πολλὰ μεθαρμόζεσθαι δυναμένων—ἀλλ᾽ αὐτοὺς τοὺς ἀριθμοὺς καὶ τὰ γεγονότα ἐφαρμόζειν ὀφείλουσι τῇ ὑποκειμένῃ τῆς ἀληθείας ὑποθέσει. Οὐ γὰρ ὑπόθεσις ἐξ ἀριθμῶν, ἀλλ᾽ ἀριθμοὶ ἐξ ὑποθέσεως, οὐδὲ Θεὸς ἐκ γεγονότων, ἀλλὰ γεγονότα ἐκ Θεοῦ· πάντα γὰρ ἐξ ἑνὸς καὶ τοῦ αὐτοῦ Θεοῦ. For my first conjecture, see 1.16.1.11, where *xxx numerus* is a duplicated translation of ἡ Τριακοντάς. There is no "number of the thirty" in Valentinian theology, but rather a Triacontad, within which are many numbers. On my second conjecture, see 1.10.3.1157. The sexual connotation of *copulare*, alluding to the Valentinian syzygies, suggests a word stronger than συνάπτειν. What I translate as "narrative," ὑπόθεσις, can equally well be translated "plot," "argument," or "doctrinal system." Irenaeus uses the same word frequently at 1.9.4 to describe the narrative structures of the *Iliad* and *Odyssey*, and to this discussion he no doubt alludes here, at 2.25.1. See Rousseau et al. 1965–1982: 1.296–299.

No other passage is as important for understanding Irenaeus' theology of arithmetic. He argues that numbers, syllables, and letters are unfit to be the foundation of a system because they are composite, they have many different qualities, and they can be used to prove anything one likes. Anyone can dream up a narrative and find numbers and letters to corroborate it. By calling numbers and letters weak, Irenaeus gives priority to Scripture and the rule of faith, which are unchanging. So the Valentinians' principal fault is in going first not to the Bible but to the world of numbers and letters. Although the Valentinians appeal to the Scriptures and to the order of the natural world, these appeals are based ultimately upon preconceived arithmetical ideas. Scripture is extra.

Irenaeus offers his own, alternative principle, captured eloquently in the ancient Latin translation *numeri ex regula*. A narrative should not take shape from numbers, but vice versa. The relationship between God and creation provides the template. Just as all things come from one and the same God, so numbers and their proper, intended use emerge from the underlying narrative of truth, the canon of faith maintained by the Church.[22]

Irenaeus' doctrine of God follows this principle. The only number symbolism he applies to God is that of the "number" one, and he grounds this whenever possible in the Bible. He says, "John preached one God Almighty and one Only Begotten Jesus Christ," in direct opposition to Valentinian interpretations of the same Gospel (1.9.2).[23] Paul's phrase "one God the Father" (Ephesians 4.6) is evidence of the Church's belief in only one God (2.2.6), a belief that has existed throughout history, from the protoplast Adam, through the prophets, to the age of the universal Church (2.9.1). That Church, he says, received a common faith in one God the Father almighty and one Jesus Christ, the Son of God (1.10.1). Elsewhere throughout *Against Heresies* he frequently insists on the unity of God.[24] As a reflection of the one God, the Church, though dispersed, dwells as if in one house, possesses one soul, and proclaims the Gospel with one mouth. Like God, the Church's tradition has a single power, and its faith is one and the same (1.10.2, 3). Despite this emphasis on divine unity, Irenaeus avoids any language that refers to God as *the* One, thus distancing himself from both Platonists and Pythagoreans.

Although Irenaeus roots his doctrine of the oneness of God in the rule of faith, he expresses it so as to refute Valentinianism. His insistence that there is but one Father responds to Valentinian claims that there are two; his

---

[22] The extended argument is at *Against Heresies* 2.15–16.

[23] *Against Heresies* 1.9.2: Τοῦ γὰρ Ἰωάννου ἕνα Θεὸν παντοκράτορα καὶ ἕνα Μονογενῆ Χριστὸν Ἰησοῦν κηρύσσοντος. Cf. *Proof of the Apostolic Preaching* 5.

[24] See e.g. *Against Heresies* 2.1.1, 2.11.1, 2.16.3, 4.38.3.

proclamation of there being only one Son counters the Valentinian notion that Only Begotten, Word, Christ, and Jesus are names for four separate entities. But Irenaeus is silent about the unity of the Holy Spirit, whose numerical integrity the Valentinians had not challenged. In Book 1, he presses home "faith in one God the Father Almighty ... and in one Jesus Christ the Son of God ... and in the Holy Spirit." Many other creeds follow this same formula, preserving intact Irenaeus' concern with the Valentinian numbering of the Father and the Son (but not the Spirit).[25]

Irenaeus' concern for Valentinianism is evident, too, in that he never uses τριάς 'trinity' of the Father, Son, and Holy Spirit, despite the use of the term in his day and despite his consistent teaching that all three are the one God.[26] To some modern readers, Irenaeus' omission of the term suggests the slow, late development of the doctrine of the Trinity. But this is to ignore that for Irenaeus τριάς, analogous to Dyad, Tetrad, or Ogdoad, was open to Valentinian overtones (e.g. 2.15.1). Further, Irenaeus says that the Valentinians, by using arithmetical terms such as 'Dyad' and 'Tetrad,' have made arithmetic a determining factor over the Pleroma. That is, they set the creation over the creator. So Irenaeus avoids the term 'Triad' of Father, Son, and Holy Spirit to ensure that his readers do not subject the divinity to mathematical abstractions.

So much for numbers in Irenaeus' theology. But what about his Bible interpretation? Irenaeus was quite attracted to the numbers found in the Bible, and he interpreted many of them symbolically. In certain respects he tried to stay close to the rule of faith; in others he was just as arbitrary as the Valentinians.

In discussing Isaiah 11.2–3 ("And the Spirit of God will rest upon him, a Spirit of wisdom and understanding ..."), Irenaeus, like the Valentinians, uses number symbolism to connect the Scripture and the natural world. According to this verse there are seven virtues that come upon the Messiah: wisdom, understanding, counsel, might, knowledge, piety, and the fear of the Spirit.[27] Irenaeus explains that the virtues refer to the seven heavens, the model Moses used for the seven-branched candlestick, in obedience to the command to fashion things as a type of what was revealed to him on the mountain (Exodus 25.40).

---

[25] Irenaeus uses this early credal practice against the Valentinians, who "confess with the tongue one God the Father," Creator of all things (4.33.3) and "one Lord Jesus Christ, the Son of God," yet they split up Christ into assorted entities. Thus, according to Irenaeus, when at church they profess faith in the unity of God and his Son they pay mere lip service. Easter creeds through the fourth century tend to follow Irenaeus. See Kelly 1950:167–204, esp. 195.

[26] See Theophilus of Antioch *To Autolycus* 2.15, dated to just after 180. On Irenaeus' Trinitarian theology, see below and, e.g., *Against Heresies* 4.38.3; *Proof of the Apostolic Preaching* 4–8.

[27] Irenaeus *Proof of the Apostolic Preaching* 9.

In another passage, Irenaeus explains some obscure numbers in the story of Jericho—why are there three spies and seven marches around the city?[28] For the explanation he draws from orthodox tradition. Rahab, who welcomed the three men spying out the entire inhabited land, reveals in herself the Father and the Son, with the Holy Spirit. The fall of Jericho indicates the seven last trumpets. Thus, the capture of Jericho symbolizes the final age of history, when salvation will belong only to those who embrace in their hearts the three divine persons.

When discussing God's command to Gideon to break up the altar of Baal and cut down the Asherah, Irenaeus interprets his taking ten men as a prophecy of Christ.[29] Irenaeus' version of Judges 6.27 must have had the alphabetic numeral for ten (ι') instead of the word δέκας, to read: καὶ ἔλαβεν Γεδεων ι' ἄνδρας. The numeral ten, an ἰῶτα, is Jesus' initial. This, Irenaeus says, shows that Gideon appeared to have Jesus as his help.

Irenaeus interprets the thirty-, sixty-, and hundredfold fruit in the parable of the sower as three levels of reward in the hereafter (5.36.2, citing Matthew 13.8). The hundred represent those who will be taken up into the heavens, the sixty, those spending time in paradise, and the thirty, those inhabiting the city (i.e. the heavenly Jerusalem). To corroborate this interpretation, Irenaeus claims as his authority the elders who were disciples of the apostles. According to them, those being saved advance by stages, by rank, through the Spirit toward the Son, and through the Son toward the Father. Thus Irenaeus draws from an oral tradition to elucidate a parable whose interpretation is not immediately clear. Unlike the Valentinians, Irenaeus pays no attention to the symbolism of the numbers themselves. Rather he focuses on their structure, that of a tripartite hierarchy.

Eschatological numbers fascinated Irenaeus. I have already mentioned how he takes the seven marches around Jericho to represent the trumpets of the end of time. His most sustained discussion of number symbolism occurs at the end of Book 5 of *Against Heresies*, where he treats numerous passages in Daniel and Revelation, whose symbolic numbers seem to have attracted as much attention then as they have since. Irenaeus suggests that, because one day is as a thousand years to the Lord, the world must come to an end after six millennia, reflecting the six days in which it was created (5.28.3, citing 2 Peter 3.8). He also addresses the very contentious issue of the interpretation of the number of the beast (Revelation 13.18). He begins by noting that the name of the beast is fittingly

---

[28] Joshua 2, Joshua 6; Irenaeus *Against Heresies* 4.20.12. Joshua 2 (MT and LXX) mention only two spies. Joshua 6.5 (LXX) speaks of only one trumpet, whereas Irenaeus speaks of seven (and he calls them "final": 1 Corinthians 15.52). Irenaeus seems to take the trumpets at Jericho to foreshadow the end-times trumpets of Revelation 8.2, 6.

[29] Judges 6.27 (LXX Vat., not Alex.); Irenaeus, fragment 18 (Harvey). The syntax of much of this passage is unintelligible, despite its attestation in three manuscripts.

666, since the number shows how he sums up in his person the pervasive spread of wickedness before the deluge (5.29.2). Noah, after all, was six hundred at the time of the flood (Genesis 7.6). And the beast, the sum of idolatry, is symbolized by Nebuchadnezzar's image, which had a height of sixty cubits and a breadth of six. Six hundred, sixty, and six make 666, QED. This recapitulation, where six recurs in units, tens, and hundreds, signifies the recapitulation of the apostasy at the beginning, middle, and end of history (5.30.1). Staying true to his principles, Irenaeus draws from the Scriptures, choosing verses whose treatment of the number six allows for a broader, moral treatment of Revelation. He does not characterize the number six as a falling short of the perfection of seven, nor does he suggest, at least here (see below), that the beast's number has anything to do with the psephic value of a name (in contrast to Marcus' 888, the psephic value of Ἰησοῦς). The first of these explanations would have been a difficult sell, since six was considered a perfect number, and it had only positive connotations in ancient number symbolism.[30] The second explanation would have veered too close to Marcus' techniques and would have encouraged speculation in gematria, which Irenaeus found exasperating.

Some Christians, probably a sizable minority, held that the number of the beast was 616.[31] Irenaeus criticizes this position, partly because 616 disrupts an intentional numerical pattern that symbolizes the recapitulation of evil, and partly because this reading depends upon textual corruption. Irenaeus says that the number results from a common error, the Greek letter *xi* unraveling so as to look like an *iōta*.[32] Those who depend upon this reading may seek a sure and certain interpretation, but in so doing they open themselves up to deception. They are working from a deficient manuscript, they have not consulted the oral tradition of the apostles, and they have ignored the proper, moral significance of 666.[33]

Although Irenaeus recognizes that the number 666 indicates the psephic value of a name in Greek, he treats it as a secondary line of interpretation. And he pleads for temperance in solving the riddle (1.30.3). He discusses several possibilities—Εὐάνθας, Λατεῖνος, and Τεῖταν—each of which adds up to 666. Τεῖταν has an added bonus: it has six letters. Despite these possibilities, Irenaeus says, we should not endanger ourselves by claiming with certainty that we know the name. If it had been imperative that the name be clearly proclaimed now, the seer would have written the name directly. In line with his general theories

---

30   *Theology of Arithmetic*, s.v.
31   Along with Irenaeus' testimony, several New Testament fragments confirm this variant reading. See Nestle et al. 2004, s.v. Rev. 13.18.
32   Or a copyist may have inserted this comment, Harvey's suspicion (1857: s.v.).
33   See Rousseau et al. 1965–1982:1.331–333.

of biblical interpretation, Irenaeus treats it as an obscure number that does not allow an immediate, obvious interpretation.[34]

Irenaeus' most famous use of number symbolism is his argument for there being four and only four Gospels (3.11.8).[35] It is not so much Bible interpretation as meta-Bible interpretation. Responding simultaneously to those who espoused more Gospels (the Valentinians) and to those who held to a much smaller number (Marcion, who accepted only Luke, and an abridged form at that), Irenaeus claims that the Gospels had to have been four, no more and no fewer. There are four regions of the world and four universal winds.[36] The Church is spread throughout the whole earth and the Gospel is the "pillar and support" (1 Timothy 3.15) of the Church and the spirit of life.[37] Because of all these things, the Church fittingly has four columns, breathing incorruption and granting people life from all directions. From this it is evident that the Word, the craftsman of all things, after manifesting itself to humanity, gave a quadriform gospel encompassed by a single Spirit. The Word rests on the cherubim and the cherubim have four faces (Psalms 79.2; Ezekiel 1.6, 1.10). These faces are images of the activity of the Son of God. Following the language and order of Revelation 4.7 (and not Ezekiel 1.10), Irenaeus interprets the lion, ox, man, and eagle as, respectively, John, Luke, Matthew, and Mark, on the basis of the opening lines of each Gospel.[38] The four animal shapes reflect the four activities of the Son of God, and the four Gospels. For the same reason, four covenants were given to humanity, the first before the deluge, the second to Noah, the third to Moses, and the fourth is that which "renews man and recapitulates in itself everything, through the gospel raising and granting wing to men for the heavenly kingdom."[39] Thus everyone who nullifies the shape of the Gospel and introduces more or fewer faces of the Gospel is foolish and ignorant (3.11.9). Those who have more (the Valentinians) claim they have found something greater than the truth. Those who have fewer (the Marcionites)

---

[34] At *Against Heresies* 2.27.1–2.28.2–3, Irenaeus says that the more veiled, opaque passages of Scripture should not be used to decipher parables. The passages that are clearest should be the interpretive key for the more obscure. That there are Scriptures we don't understand is to be expected. After all, we do not understand many things in the natural world.

[35] Cf. an Armenian abridgement, fragment 6 at Ter Mĕkĕrttschian and Wilson 1919:737. Irenaeus' is the first attested argument for the four canonical Gospels, but we should not infer from this that it is his invention. Irenaeus regularly reproduces, even verbatim, the thought of his predecessors without attribution. See Hill 2006.

[36] Cf. *Theology of Arithmetic* 24.10, 29.15.

[37] Πνεῦμα ζωῆς. The pun and intent in relating the Scripture to world geography is better seen by translating the phrase "wind of life."

[38] For the larger patristic tradition see Stevenson 2001.

[39] τετάρτη δὲ ἡ ἀνακαινίζουσα τὸν ἄνθρωπον καὶ ἀνακεφαλαιοῦσα τὰ πάντα εἰς ἑαυτήν, ἡ διὰ τοῦ εὐαγγελίου ἀνιστάσα καὶ ἀναπτεροῦσα τοὺς ἀνθρώπους εἰς τὴν οὐράνιον βασιλείαν.

nullify the dispensation of God. God creates all things to be harmonious and well fitted, so the form of the gospel too had to be harmonious and well fitted. Thus the four gospels alone are true and certain, and admit neither increase nor decrease.

How well does this argument for the four Gospels conform to Irenaeus' principles? He does not seem to appeal to the rule of faith, and it looks like the argument arises from a predilection for number symbolism. Is Irenaeus committing the Valentinians' sin of *regula ex numeris*?

Consistency in Irenaeus' Thought

Irenaeus claims that the Church's teachings, unlike those of the Valentinians, are well fitted (2.15.3). He calls the Church's proclamation a rhythm, fitted to the things that have been created by the rhythm (*apta est enim haec rhythmizatio his quae facta sunt huic rhythmizationi*). That is, the rule of faith conforms exactly to the contours of creation because it is the very rule by which creation was shaped. Creation is well ordered, and the tradition fits the order of creation.[40] The sentiment is less an argument than a pair of self-referential claims, akin to the early Christian notion that by the Word all things were made, and that this very Word is that which the Church proclaims.[41] The claims are two arcs of a single circle. The causes underlying the structure of the world reside within the Church, and the Church's proclamation is made manifest in the structures of the world. Throughout *Against Heresies*, Irenaeus expounds the rule of faith and emphasizes this internal consistency. He is no epistemological foundationalist.

But is Irenaeus consistent? How well fitted is his rule of truth to the principles he outlines? He charges the Valentinians with mishandling Scripture, with putting numbers ahead of doctrine. But has he himself committed this very error? Recall the four main lines of Irenaeus' substantive critique of Valentinian number symbolism, but phrased as principles that Irenaeus must necessarily defend. First, the numbers in one's taxonomy of the godhead should be consistent. Second, theology should not derive from and depend upon the changing linguistic or mathematical habits of a particular society. Third, numerical patterns of the natural world, if used to justify one's theology, should be applied consistently. Fourth, numbers in theology should emerge from the entire body of Scripture, and proof texts should be used with regard to their context. This last principle is especially important for him: numbers should emerge from

---

[40]   Cf. Perkins 1992:279.
[41]   John 1.3 and, e.g., 1 Thessalonians 2.13.

the rule of faith (the Scripture being part of that rule), not the other way around.

In his second and third principles Irenaeus is inconsistent. He criticizes exegetical techniques that he himself uses. When he says that Gideon took 10 men as a prophecy of Jesus, he breaks a rule similar to the one he accuses Marcus of breaking. It was known in the late second century, as today, that habits of numeration in Gideon's time were quite different than those in the second century.[42] So Irenaeus' anachronistic use of a contemporary Greek technique to interpret Judges 6.27 is just as misplaced as Marcus' application of Greek psephy to what were originally Hebrew names. Perhaps Irenaeus was thinking in this instance of the Hebrew alphabet, and treating Gideon's 10 men as a *yod*, which, like the *iōta*, stood for the number ten. But this too would have been anachronistic, since alphabetic numeration had been in wide use in Hebrew only for about 120 years.[43]

As for the third principle, the need for one's number symbolism to be consistent with nature, Irenaeus is unfair to his opponents. In Irenaeus' day, the four winds and the four cardinal points were unquestionable, natural phenomena subject to neither change nor social convention. Even if we grant him the science, was it any reason to conclude that there are four and only four Gospels? Recall his criticizing the Valentinians for justifying their various levels of the Pleroma on the basis of inappropriate divisions of time. Irenaeus is just as arbitrary. A Valentinian, for instance, could have argued for the five Gospels—Matthew, Mark, Luke, John, and the Gospel of Truth—on the basis of the five senses, the same basis upon which deutero-Simon argues for the perfection of the Pentateuch. Whose argument is stronger? Are the Gospels more like geography or sensation? And if the Gospels resemble the four winds, as Irenaeus claims, then how does the number of epistles or other biblical books fit into that analogy? Why the specific number of letters by Paul and other apostles? Why the number of Old Testament books? If the Gospels represent the four winds, what do the other books represent? After all, Irenaeus insists that if the Valentinians find evidence for the Dodecad in the calendar year, then they must find chronological divisions that are prototypes of the Decad and Ogdoad as well, to be consistent. By the same token, the numbers of the other books of the Bible should be reflected in the appropriate parts of the natural world. Both Irenaeus and Valentinians depended upon incomplete, ad hoc natural analogies.

---

[42] See e.g. Aelius Herodianus (fl. second c. CE), Περὶ ἀριθμῶν (TLG no. 87.42).

[43] Save for a rare early example—a coin from the reign of Alexander Jannaeus (103–76 BCE)—Hebrew alphabetic numerals are known to have been widely used only after 66 CE; Lieberman 1987:193–198.

In the first and fourth principles, however, Irenaeus is entirely consistent. He keeps numbers subservient to God, their creator, and interprets numbers in conformity with the rule of faith preserved in the churches of Asia Minor, Gaul, and Rome. He is especially careful to preserve inviolate the first principle, that of a consistently numbered godhead. In pursuing his biblical exegesis, Irenaeus tries to apply the fourth principle, making his interpretation come from the Scriptures and the rule of faith itself, or at least remain in conformity with them.

Irenaeus held to four Gospels not because of the weather patterns but because of the churches' tradition (*Against Heresies* 3.1). He learned from the elders of Asia Minor, who themselves preserved what the apostles taught them, that there are but four Gospels. This is the apostolic rule of faith, and Irenaeus treats it as having the authority of Scripture. The rule of faith preserves four and only four books, and this unalterable fact enlightens other, more obscure parts of the tradition. The four Gospels explain the meaning of the four faces of the cherubim, as well as of the four covenants God gave throughout human history. Such numbers are drawn exclusively from the tradition.

We might imagine the Valentinians' objecting, claiming that they too were working from within a tradition, albeit a specially revealed one. Irenaeus' counterargument, however, would be strong. Doctrine respects the narrative structures of the Scriptures, and it engages in the entire breadth of the tradition. It flows out of the tradition, not into it. The tradition does not depend on any one person, school, movement, or special revelation, but is the possession of all the churches. The Valentinians cannot justifiably claim to be following this rule. If they could, they would be able to point everywhere to churches that preserve their tradition from the apostles (3.1–4), they would have a place for the sacred number five, and their system would harmonize with the Scriptures taken in their broader context.

Irenaeus' fourth principle is his chief axiom. The rule of faith is a constant, consisting of the Scriptures, the oral tradition, and the life and teaching of the churches founded by the apostles. Numbers should be treated as a part of that rule, and numbers from another rule should not force their way in. For Irenaeus, the various numbers in Scripture are not so much proofs as implications of his system. The four animals of Ezekiel 1 and Revelation 4 do not justify the claims that there are exactly four Gospels; rather, these verses are explained by that part of the tradition. Irenaeus' phrases of inference—"for also" (καὶ γάρ), "and because of this" (καὶ διὰ τοῦτο)—do not look backward, to a basis for belief, but forward, to its implication. Throughout *Against Heresies*, Irenaeus uses inferential language for two purposes: one for proof, and the other to show off a doctrine's explanatory power. When interpreting the four Gospels, he uses only the second technique. His clause of inference "for since" (ἐπεὶ γάρ; 3.11.8.176–177) explicates rather than justifies. He is, in the end, not proving the number of Gospels but rather explaining the coherence of that number. He cloaks his explanation in clauses of inference to strengthen the persuasive power of his rhetoric, albeit at the expense of clarity.

Irenaeus responded thoroughly to the Valentinian theology of arithmetic by setting down strictures and principles that applied to all of Christian theology and exegesis. He carefully upheld the principles he most cared about. In his exegesis, he maintained his commitment to orthodoxy as a driving principle, but he used some of the same arithmological techniques the Valentinians, Philo, and Plutarch used. Irenaeus was typical of many orthodox writers in his day, but his approach, with its ideals and inconcinnities, is only one example of the orthodox approach to number symbolism. A more complete picture requires us to consider Irenaeus' near contemporary, Clement of Alexandria.

7

The Orthodox Possibilities of the Theology of Arithmetic

Clement of Alexandria

CLEMENT, A CHRISTIAN INTELLECTUAL who flourished in late second-century Alexandria, offers a perspective on the orthodox theology of arithmetic that departs from, yet complements, that of Irenaeus. Unlike Irenaeus and his head-on refutation, Clement criticizes the Valentinians subtly, preferring to co-opt heretical number symbolism for orthodox ends. And Clement's interest only begins with gnosticizing Christians. Just as important to him are Platonists and Stoics, whom he engages with number symbolism to show that Christ and Christian theology surpass their teachers and doctrines.

In this chapter I explore three significant arithmological motifs found in Clement's writings. In each case, Clement engages an area previously discussed by Irenaeus, a Valentinian, or a school of philosophy. First is his doctrine of God, and his use of the epithet 'One' for the godhead, a point of concern to Irenaeus and Platonists. Second is Clement's use of number symbolism to build a novel anthropology that he intended to be a viable alternative to Valentinian and Stoic schemes. And third is Clement's interpretation of the number symbolism built into the story of the Transfiguration, an allegory that departs from both Marcus' version and Irenaeus' dissent. In each of these, Clement threads the theoretical needle between or around the two alternatives, showing how an orthodox theology of arithmetic can remain both faithful to the tradition and compelling to the cultural elite.

Numbers in Clement's Doctrine of God

Very little is known of Clement. The wealth and social status of his family allowed him to travel to pursue his education. Clement credits with his training in Christianity teachers from across the Roman world—Greece, Italy, Lebanon,

Syria, and Egypt—particularly a certain Pantaenus, whose missionary travels to India are mentioned by Eusebius.[1] Once he settled in Alexandria, Clement conducted what was probably an informal Christian school, not to be confused with the famous academy that started under the auspices of Origen.[2] After the persecution of Christians in 202 Clement left Alexandria, and died presumably a short time later.[3]

Several of Clement's numerous writings survive, in whole or in part, and in various states of editorial polish.[4] Evident in all his writings is an unflagging commitment to Christian orthodoxy. His style and spirituality differ from those of his contemporaries Irenaeus and Tertullian, but he nevertheless identifies himself with their opinions and the traditions preserved by the Church. Like them, he holds to a single God, the Father, and a single Son, the express image of the Father. As a self-proclaimed "ecclesiastic," he uses the various titles for God and his Word—e.g. 'Son of God,' 'Christ,' 'Savior,' 'Instructor,' and 'Jesus'—to describe one and the same person, unlike the Valentinians.[5] He holds to no theology of emanations from God, so he has no system of aeons organized into numerically symbolic groups.[6]

But Clement is more adventurous than other orthodox writers in using arithmetic to describe the movement from the godhead to the structures of creation. In his view, below the Father and the Son are numerous beings that form an elaborate hierarchy extending from heaven to earth, all with their source of being in unity, and unity as their goal. Clement says that these lower beings save each other "from One and through One."[7] "From One" refers to the Father; "through One" to the Son. Abraham, according to Clement, embraced this unity of God in the alteration of his name from Abram, since the *alpha* that was inserted into his name represents his knowledge of the one and only God.[8] Clement's sense of the unity of God is so strong that he calls him "the One" and applies the attributes of the number one to the godhead: just like the number one, God "the One" is indivisible, and therefore infinite, realized in his lack of extension.[9]

---

[1] Clement of Alexandria *Stromateis* 1.1.11; Eusebius *Church History* 5.10–11.
[2] Jakab 2001:93–106.
[3] For more on Clement's life, see DECL, s.v., and works cited there; on his corpus and his theological thought, see Quasten 1953:5–36, with updated bibliography in TRE, s.v., and DECL, s.v.
[4] For a fresh perspective on Clement's corpus, see Bucur 2009.
[5] Clement prefers to distinguish himself and other orthodox churchmen as "of the Church," a description he denies to Valentinians and other opponents who cultivate private revelation. See Kovacs 1997:415n5.
[6] On the differences and similarities between Clement and the Valentinians, see Davison 1983 and Edwards 2000.
[7] *Stromateis* 7.2.9.3.
[8] *Stromateis* 5.1.8.6.
[9] *Stromateis* 5.12.81.6. For discussions of the philosophical dimensions of Clement's use of 'one,' see Choufrine 2002:165–166, 174–175, 186–188.

Clement even says that God "calls himself one," on the basis of John 17.21–23: "In order that all might be one, just as you, Father, are in me and I in you," and so forth.[10] But lest it be thought that his God is the Platonist One, Clement interprets John so as to affirm that God transcends all number: "God is one, and beyond the One, and above the Monad itself."[11] Adopting here the now familiar idea that the Monad transcends the Hen, Clement says that God stands not only above the One (the Platonists' highest principle) but above the Monad itself (the Pythagoreans'), and is therefore beholden to no number. Clement cautions readers that the epithet 'One' for God is an approximation, and not a true predicate of He Who Cannot Be Named.[12] Such negative theology was standard in the Platonism of his day. But Clement presents this idea of the indescribability of God, an idea that becomes quite important in Plotinus' writings, not to press Christianity into a Platonic mold but to reinforce belief in the transcendence of God. For Clement there is no category, including number, that comprehends and stands over his nature. Thus Clement's metaphors and pedagogical tools may be philosophical in origin, but in the substance of his theology he stands with Irenaeus as a Christian monotheist, not a Platonist.[13]

Although God stands above arithmetic, Clement finds arithmetical unity a helpful metaphor of the divine, and he states that man's goal is a similar kind of unity. As a person becomes divinized and moves into a state of dispassion, he becomes purely "Monadic."[14] This unity is epitomized for Clement in the Church: "For just as God is one and the Lord is one ... that which is most highly treasured is praised for its solitude since it is an imitation of one principle. Thus, the one Church also has a portion in the nature of the One, which nature the [heretics] strive to chop into many heresies."[15] This joint share in God's unity allows the Church to collect people "into the unity of the one faith of its proper testaments—rather of the one testament from different ages—by the will of the one God, through the one Lord."[16] Thus the Church, which is the earthly image of the heavenly Church, reflects precisely the unity of God, and humanity's return to that unity.[17]

---

[10]   Clement of Alexandria *Instructor* 1.8.71.1.

[11]   ἓν δὲ ὁ θεὸς καὶ ἐπέκεινα τοῦ ἑνὸς καὶ ὑπὲρ αὐτὴν μονάδα. *Instructor* 1.8.71.2. See p. 181 below.

[12]   *Stromateis* 5.12.82.1.

[13]   This notion holds for all the Church Fathers, often wrongly characterized as Christianizing Platonists. See Edwards 2002 and Ramelli 2011.

[14]   *Stromateis* 4.23.152.1.

[15]   ἑνὸς γὰρ ὄντος τοῦ θεοῦ καὶ ἑνὸς τοῦ κυρίου, διὰ τοῦτο καὶ τὸ ἄκρως τίμιον κατὰ τὴν μόνωσιν ἐπαινεῖται, μίμημα ὂν ἀρχῆς τῆς μιᾶς. τῇ γοῦν τοῦ ἑνὸς φύσει συγκληροῦται ἐκκλησία ἡ μία, ἣν εἰς πολλὰς κατατέμνειν βιάζονται αἱρέσεις· *Stromateis* 7.17.107.4.

[16]   "εἰς ἑνότητα πίστεως" μιᾶς, τῆς κατὰ τὰς οἰκείας διαθήκης, μᾶλλον δὲ κατὰ τὴν διαθήκην τὴν μίαν διαφόροις τοῖς χρόνοις, ἑνὸς τοῦ θεοῦ τῷ βουλήματι δι᾽ ἑνὸς τοῦ κυρίου συνάγουσαν τοὺς ἤδη κατατεταγμένους· *Stromateis* 7.17.107.5.

[17]   *Stromateis* 4.8.66.1.

The Anthropological Decalogue

Clement's longest extant work is his *Stromateis* (*Miscellanies*), a lengthy patchwork of discussions on various theological topics, intended for study by Christians prepared to think about advanced topics. Book 6 of the *Stromateis* contains his richest meditation involving number symbolism (§§133–148). Following up a principle mentioned earlier in Book 6, that Christians can profitably use the four mathematical disciplines, Clement applies arithmetic to Scripture, particularly the Ten Commandments, and Scripture to arithmetic. In his excursus, a small treatise in its own right—I call it *On the Decalogue* here for convenience—Clement demonstrates how an advanced Christian might use his education, especially in arithmetic or geometry, to explain passages from the Bible.

On the Decalogue falls in two parts. The first half (133.1–137.1) relates Moses' Decalogue and its number symbolism to creation and anthropology. Clement uses the number ten as a structural device around which to frame a Christian alternative to Stoic anthropology, and his system, difficult to understand on first reading, illustrates how a Christian might profitably use number symbolism. In the second half (137.2–148.4), Clement explains the theological significance of eight of the Ten Commandments. When he arrives at the commandment to keep the Sabbath holy, Clement pursues a lengthy tangent, to discuss the relationship between the numbers six, seven, and eight (138.5). He interrupts this tangent—which draws from Jewish, Christian, and Hellenistic arithmology—with yet another, an arithmological interpretation of the Transfiguration (140.3), the centerpiece of *On the Decalogue*, which I explore toward the end of this chapter.

The first half of *On the Decalogue* begins with a meditation on the well-ordered multiplicity of creation and humanity. Although not a unity, as God is, says Clement, creation has been imprinted with a unifying harmony that reflects its divine origin. To demonstrate this harmony, he carefully constructs three lists of ten elements each—types of decalogues, all of which, he says, the Decalogue encompasses (περιέχει: 133.3, 4). The first, the heavenly decalogue, is sun, moon, stars, clouds, light, wind, water, air, shadow, and fire. The second decalogue's ten items relate to earth and the sea: humans, cattle, reptiles, beasts, fish, whales, carnivorous birds, birds of a delicate palate, fruit-bearing plants, and plants with no fruit. The first decalogue does not seem to follow a specific order, but the second list does. It follows in reverse the days of creation found in Genesis 1, which goes from plants (day three), to reptiles, birds, and fish (day five), to quadrupeds, land reptiles, beasts, and humans (day six).[18] The

---

[18] Possibly the first decalogue, too, follows the order opposite from Genesis. But there are notice-able differences: fire and clouds are not mentioned until Genesis 11.3 and 9.13, respectively, and ἀήρ does not feature in LXX Genesis at all. These three items excepted, the order in Genesis

rough correspondence shows that Clement shaped the Genesis account to suit his number symbolism. In this he resembles the Valentinians, who drew from Genesis lists of ten and twelve items to support their doctrine of the Decad and Dodecad.[19]

The third decalogue Clement presents (134.2) is that of the human being, who he says consists of the five senses, along with (6) the ability to speak, (7) the ability to generate, (8) the formed spirit, (9) the ruling faculty of the soul, and (10) the characteristic mark of the Holy Spirit (a mark applied through faith).[20] This anthropological decalogue features in Clement's other writings as well, and provides an important contrast with similar structures in Valentinianism and Stoicism. Details in his use of this decalogue illustrate how he used numbers to articulate highly nuanced points about the creation and salvation of the world. Clement may have imitated Valentinians and Stoics, but that was to entice them to embrace orthodox Christianity.

For example, in Book 2 of *Stromateis*, in his interpretation of Exodus 16.36 ("The omer was the tenth of the three measures"), repeating Philo nearly verbatim, Clement identifies the three measures as sense perception, reason, and intelligence, as well as the intended objects of these three faculties.[21] He then adds his own thoughts to Philo's explanation: the true and just measure is the gospel teaching that "it is not what enters into the mouth that defiles a person, rather, that which exits a person's mouth is that which defiles a person."[22] Clement now draws from his anthropological decalogue.[23] That same measure is the "decad that encompasses the human being." The three measures of Exodus 16.36 allude in summary form to that decad.[24] He then explains the decad in terms similar to those found in *On the Decalogue*: "That might be both the body and the soul: the five senses, the vocal faculty, the generative faculty, and the faculty of under-standing or of the spirit, or whatever you want to call it."[25]

---

would be shadow, wind (= spirit), water, light, sun, moon, and stars. Possibly Clement considered fire and air to be implied in the first day of creation, and clouds in the second. If so, then his heavenly decalogue, like the earthly one, follows the days of creation in reverse order.

[19] See p. 46 above.

[20] τὸ διὰ τῆς πίστεως προσγινόμενον ἁγίου πνεύματος χαρακτηριστικὸν ἰδίωμα. The eighth, the formed spirit, is literally "that which is spiritual/breathing according to the formation" (τὸ κατὰ τὴν πλάσιν πνευματικόν), discussed below.

[21] Philo *On the Preliminary Studies* 100.

[22] οὐ τὰ εἰσερχόμενα εἰς τὸ στόμα κοινοῖ τὸν ἄνθρωπον, ἀλλὰ τὰ ἐξερχόμενα διὰ τοῦ στόματος ἐκεῖνα κοινοῖ τὸν ἄνθρωπον, Clement's rendering of Matthew 15.17–18, at *Stromateis* 2.11.50.2–3.

[23] *Stromateis* 2.11.50.1.

[24] *Stromateis* 2.11.50.3: τοῦτ', οἶμαι, τὸ κατὰ θεὸν ἀληθινὸν καὶ δίκαιον μέτρον, ᾧ μετρεῖται τὰ μετρούμενα, ἡ (Mondésert 1954; Stählin incorrectly writes ἥ) τὸν ἄνθρωπον συνέχουσα δεκάς, ἣν ἐπὶ κεφαλαίων τὰ προειρημένα τρία ἐδήλωσεν μέτρα.

[25] *Stromateis* 2.11.50.4: εἴη δ' ἂν σῶμά τε καὶ ψυχὴ αἵ τε πέντε αἰσθήσεις καὶ τὸ φωνητικὸν καὶ σπερματικὸν καὶ τὸ διανοητικὸν ἢ πνευματικὸν ἢ ὅπως καὶ βούλει καλεῖν.

Clement then says one must overleap all these faculties so as to stand at the mind, as if overleaping the nine portions of the universe. These, he says, are the four elements (that is, earth and the sublunary region; portion 1), the seven planets (portions 2–8), and the 'unmoved ninth' (the fixed sphere of stars: portion 9). What he has just termed the mind—the tenth portion, the complete number—resides above these nine, and is one's attainment of the knowledge of God.[26] Thus, the nine faculties of the human being, capped by the tenth, the mind, resemble the structure of the universe, in which nine celestial levels are subordinate to God as the tenth.[27] Clement's cosmological decalogue depends directly upon Philo, but his anthropological one is original.[28]

Back to Book 6 and *On the Decalogue*. Clement twice identifies the mark of the Holy Spirit explicitly as the tenth number or element in the human being (134.2: δέκατον; 135.1: τὸν δέκατον ἀριθμόν), thus emphasizing afresh his tenfold anthropology. This point is easily missed when reading 135.1. This terse passage describes the ninth and tenth human elements (ruling faculty and the characteristic mark of the Holy Spirit) as agents that perfect other activities. The text is difficult to translate without some expansion:

> The soul is added [to the senses and limbs]. And the ruling faculty, by which we reason and which is begotten without the casting of seed [cf. Hebrews 11.11], is added before this, so that the tenth number [i.e. the characteristic mark of the Holy Spirit] is brought in without it [i.e. seed]. By these things [i.e. the ruling faculty and the characteristic mark of the Holy Spirit] every activity of a person is perfected.[29]

In this terse passage (made difficult by the vagueness of the relative pronoun ὧν) Clement notes that the ninth and tenth anthropological elements—the ruling faculty and the characteristic mark of the Holy Spirit—are not dependent

---

[26] *Stromateis* 2.11.50.4–2.11.51.1.

[27] At 2.11.50.4 (see the Greek text above), there is some admitted confusion. The five senses plus what seems like three other faculties adds to eight. The wording seems to suggest that διανοητικόν and πνευματικόν are equivalent terms for the eighth faculty. At 2.11.51.1 the mind is called the tenth faculty, in analogy to Philo's cosmological decalogue. Where is the ninth? At *Stromateis* 6.16.134.2 the faculties of understanding and of the spirit are the eighth and ninth levels. Clement's offhand remark, at 2.11.50.4, "or whatever you want to call it," suggests that there was either terminological fluidity or, as I argue below, a difference of opinion between those who wished to subordinate the spiritual faculty to the faculty of understanding and rank them eighth and ninth respectively, and those who preferred the opposite.

[28] Philo *On the Preliminary Studies* 102–106.

[29] ἐπεισκρίνεται δὲ ἡ ψυχή. Καὶ προεισκρίνεται [Stählin 1909: προσεισκρίνεται] τὸ ἡγεμονικόν, ᾧ διαλογιζόμεθα, οὐ κατὰ τὴν τοῦ σπέρματος καταβολὴν γεννώμενον, ὡς συνάγεσθαι καὶ ἄνευ τούτου τὸν δέκατον ἀριθμόν, δι᾽ ὧν ἡ πᾶσα ἐνέργεια τοῦ ἀνθρώπου ἐπιτελεῖται.

upon physical generation but are bestowed from above. They are, together, agents of perfection. These two highest faculties stand above and apart from the lower eight.

This is not Clement's only anthropological decalogue. He presents another one, independent from but compatible with his main decalogue, at 134.3, where he states that the law was laid down for the ten parts of the human, and he restricts the list to the five pairs of sense organs—sight, hearing, smell, touch, and taste. The doubled sense organs resemble the two tablets upon which the Decalogue was inscribed. And the feet and hands also resonate with the Decalogue. Clement says, in a rather confusing formulation, "Again, the laying of the law seems to be assigned to these certain ten human parts: to sight and hearing, to smell, and to touch and taste and to their assisting organs, being double, both to hands and feet."[30]

His major and his minor anthropological decalogues show that Clement saw the human being, both specifically (body only) and generally (from lowest faculty to highest), as being imprinted with the decad. These patterned decads are not unusual. The symbolism of the number ten runs throughout Clement's works. Later in this excursus (145.7) he notes that the Decalogue—thanks to the *iōta*, the Greek numeral for ten—invokes the blessed name 'Jesus.'[31] Clement earlier (84.5) calls the number ten "all perfect."[32] In Book 2, he likens the number ten to attaining the knowledge of God. For Clement, this level of perfection explains why a tithe (and no other denominator) was to be given to God, and why the Paschal feast starts on the tenth of the month.[33]

---

[30] A difficult passage: ἔτι πρὸς τούτοις δέκα τισὶν ἀνθρωπείοις μέρεσι προστάσσειν ἡ νομοθεσία φαίνεται. τῇ τε ὁράσει καὶ ἀκοῇ καὶ τῇ ὀσφρήσει ἀφῇ τε καὶ γεύσει καὶ τοῖς τούτων ὑπουργοῖς ὀργάνοις δισσοῖς οὖσι, χερσί τε καὶ ποσίν. Does this mean the five senses plus the four limbs? Or does it mean each of the five sense-perceptive faculties understood "doubly," with the hands and feet a later scribal gloss? Under the first option, the total comes to nine, not ten. The second alternative has problems too. Even though there are two eyes, two ears, and two nostrils, it is unclear what pairs of sense organs belong to touch and taste. Possibly Clement has garbled the Valentinian account, where the four senses of sight, hearing, smell, and taste (divided into bitter and sweet) have two organs each, an image of the upper Ogdoad. See Irenaeus *Against Heresies* 1.18.1 and discussion at p. 52n69 above. Compare the *Theology of Arithmetic* 68.3, where there are said to be seven orifices in the head, probably counting the tongue singly. Maybe it is the symmetry (and therefore doubleness) of the sense faculties that is key here. Note, too, that Clement's order of the senses differs (as does Plutarch's: *The E at Delphi* 12 [390b]) from that found in the *Paraphrase of the "Apophasis Megale."* See p. 100n53 above. A third interpretation, which supplements the second, seems most likely: Clement is referring to the ten toes and ten fingers. Thus, he is stating here not one but three anthropological decalogues: the senses (doubled), the fingers, and the toes.

[31] See pp. 46, 76, 118 above.

[32] ἡ δεκὰς δὲ ὁμολογεῖται παντέλειος εἶναι. See p. 54n78 above.

[33] 2.11.51.1–2. Tithe: Exodus 29.40, Leviticus 6.20; Pascha: Exodus 12.3 (see p. 90 above).

The broader parallels between Clement and Monoïmus are noteworthy. Both authors plumb the books of Moses to locate patterns and groups of ten. Both play on the role of *iōta* as a Greek numeral. Yet they approach the matter very differently. Monoïmus begins with the glyph ι as a symbol of the relationship that holds between the two supreme beings and their emanating powers. Clement is interested in ten *qua* number or *qua* numeral, but not *qua* glyph. Further, he does not share Monoïmus' metaphysics, so he uses ten as a symbol mainly of the structures of creation and their interrelated connection to Scripture, not of the godhead. And unlike Monoïmus, who never uses the name 'Jesus,' Clement sees the *iōta* in the Decalogue and the Psalms as prefiguring only Jesus.[34]

A closer parallel, one pertaining to anthropology, is found in Heracleon, his contemporary and rival. Heracleon was said to be a Valentinian, and he flourished, at least for a time, in Alexandria.[35] He shared with Clement an interest in number symbolism. Relevant here are Heracleon's comments on John 2.20 ("The Jews said, 'This temple was built in forty-six years ...'"), which outline an anthropological decalogue comparable to Clement's.[36] Claiming that the temple is an image of the Savior, Heracleon analyzes the constituent parts of the number forty-six. He says that six refers to matter or substance (ὕλη), that is, the formation of man (πλάσμα). Forty, "which is the Tetrad ... which does not admit union," refers "to the infusion, and the seed in the infusion."[37] That is, Heracleon relates the number forty-six to anthropology, treating six as the lower, material number, and forty (a number he tacitly converts to the Valentinian term 'Tetrad') as the higher, spiritual number.

Heracleon shares with Ptolemy, his fellow Valentinian, a common vocabulary and an express interest in the number ten. Heracleon's term for "does not admit union," ἀπρόσπλοκος, is reminiscent of the Ptolemy's three uses of ἀσύμπλοκος 'not woven' in his *Letter to Flora* as a quasi-technical epithet for

---

[34] Clement of Alexandria *Instructor* 2.4.

[35] Clement calls Heracleon the "most approved of the school of Valentinus" (ὁ τῆς Οὐαλεντίνου σχολῆς δοκιμώτατος; *Stromateis* 4.9.71.1). For a précis of the life and works of Ptolemy and Heracleon, see Thomassen 2006:494–496. There are two very different assessments of Heracleon: Castellano 1998 and Wucherpfennig 2002. Based on the fragments, Castellano argues for and explores Heracleon's Valentinian connections. Wucherpfennig argues that Heracleon was not a proponent of gnosis, and not even a Valentinian. Much of this argument rests on the seeming lack of overt Valentinian doctrine in the extant fragments. But Castellano's research, apparently unknown to Wucherpfennig, identifies convincing Valentinian themes. Michael Kaler's observation (pers. comm.), however, that Origen never calls Heracleon a Valentinian suggests that research on Heracleon is yet in its early stages.

[36] Heracleon, fragment 16 (= Origen *Commentary on John* 10.38.261). For other aspects of Heracleon's number symbolism see p. 138 below.

[37] Heracleon, fragment 16 (trans. Heine 1989–1993): "ὃ τετρὰς ἐστίν," φησίν, "ἡ ἀπρόσπλοκος," εἰς τὸ ἐμφύσημα καὶ τὸ ἐν τῷ ἐμφυσήματι σπέρμα.

the Decalogue, the most perfect of the three parts of the law.[38] To Ptolemy, the Decalogue is "not interwoven" with evil, and is therefore at the highest of the three levels of the Law, which suggests that Heracleon regarded the "unionless Tetrad" as the highest principle, corresponding to the place Ptolemy assigns the Decalogue.[39]

Heracleon identifies the Tetrad with the infusion and its seed, "infusion" (ἐμφύσημα) here being another technical term commonly used by Christian authors to refer to God's breathing life into Adam.[40] Heracleon has already invoked the language of Genesis by relating the number six to the material creation of man. By mentioning the infusion and the seed in the infusion (a differentiation made but not explained), Heracleon alludes to the story of God's breathing into Adam, and interprets the account as the implanting of the Tetrad in the human race. The forty-six years of the temple are an image of the Savior, who is himself the perfect synthesis of the Tetrad and the number six, the two parts of human nature, his higher infusion and his lowly matter. So Heracleon, like Clement, seems to have embraced an anthropological decalogue. Heracleon numbers the higher and lower faculties at four and six, whereas Clement divides the human being into two higher faculties and eight lower.

Whether Clement knows of, and now answers in his own orthodox fashion, the anthropological decalogue Heracleon based on John 2.20, cannot be determined. The opposite is just as plausible, that Heracleon dissents from Clement. But clearly, both were responding to Stoic anthropology. Stoicism would have been the basis for any Christian anthropology—whether orthodox or Valentinian—that carefully enumerated the parts of the human being. The earliest Stoics, to show the essential material unity of the human soul, rejected the tripartite soul taught by the Platonists, and instead divided the soul into a hierarchy of parts.[41] A number of these lists are extant, and a few of them differ, both in terminology and in the order and number of the parts. But the

---

[38]   Ptolemy *Letter to Flora* 33.5.1, 33.5.4, 33.6.6. See also Wucherpfennig 2002:85–86 and Sagnard 1947:654–655. Other aspects of Ptolemy's number symbolism are discussed at p. 31 above.

[39]   In this, Heracleon resembles Marcus and the renowned Valentinian teacher discussed at Irenaeus *Against Heresies* 1.11.3. See pp. 31–33 and 75n25 above.

[40]   See e.g. Irenaeus *Against Heresies* 1.5.6, 1.30.9, 1.30.14; Justin Martyr *Dialogue with Trypho* 40.1; Clement of Alexandria *Instructor* 1.3.7.3, *Epitomes* 3.55.2.

[41]   On the rejection of the Platonic tripartite soul see Chrysippus *Fragment* 829 (SVF 2:226 = Origen *Against Celsus* 5.47). But in late antiquity some Stoics began to accept other ways of dividing the soul, such as the Platonic. On variations in the Stoic tradition see Schindler 1934:326–345, 53–70; van Straaten 1946:119–129; Spanneut 1957:96; and Pohlenz 1970–1972:2.100–112. The Stoic variations are dependent almost wholly upon Tertullian *On the Soul* 14, whose intention to make the Stoics contradict each other probably skews the picture. See also Dillon (1996:174–175), who lists the various divisions of the soul—sometimes contradictory systems—taught by Philo. This difficulty shows that Philo was not confused or capricious, but was aware that "each of these divisions expresses some aspect of the truth" (Dillon 1996:175).

sources generally agree that to the Stoics the human being had eight parts. The seven lower faculties—that is, the five senses, the voice, and the reproductive capacity—were said to be governed by the higher one, the ruling faculty (ἡγεμονικόν).[42]

Clement expands the eight Stoic parts of the soul into a decalogue through two modifications. He moves the ruling faculty from the eighth to the ninth place and then inserts two new faculties: the formed spirit, in the eighth position, and the mark of the Holy Spirit, in the tenth. To understand the significance of these modifications, we must understand Clement's ideas behind his two new faculties, the formed spirit (eighth) and the mark of the Holy Spirit (tenth).

Both new faculties are mentioned in *Stromateis* Book 4, where Clement discusses the potential for a gnostic—his preferred term for the spiritually advanced orthodox Christian—to become a god. He contrasts the composition of the human being in general with that of specific individuals. "So the human being in general is formed (πλάσσεται) in accordance with the form (ἰδέαν) of the connate spirit."[43] He associates this spirit with the shape of the human being, both in essence and in physical form, and he says it explains why man lacks neither form nor shape "in the factory of nature." Thus, the "connate spirit" here corresponds to the eighth faculty in his anthropological decalogue, since it is higher than, but nevertheless affects, the physical shape of human beings. In contrast with the general human being, "the individual man is characterized (χαρακτηρίζεται) by the impression (τύπωσιν) of his choices entering into (ἐγγινομένην) the soul."[44] The parallel terminology to 134.2 (quoted above: προσγινόμενον, χαρακτηριστικόν), where the tenth part, the mark of the Holy Spirit, is discussed, shows that the faculty that "impress[es one's] choices" in the soul in Book 4 is the same as the tenth element in Clement's anthropological decalogue. This is confirmed further on: "By this [impression] we say that

---

[42] Sometimes discrepancies in the sequence occur in the same source: Diogenes Laertius *Lives of the Philosophers* 7.110 (= Chrysippus, fragment 828 [SVF 2:226]) contrasts with 7.157. The primary sources that attest to the canonical eight parts are numerous: Zeno, fragment 143 (SVF 1:39 = Nemesius *On the Nature of Man* 96); Chrysippus, fragment 827 (SVF 2:226 = Aetius *Placita* 4.4.4); idem, fragment 830 (SVF 2:226 = Porphyry *On the Soul* in Stobaeus 1.49.25a); idem, fragment 831 (SVF 2:226 = Iamblichus *On the Soul* in Stobaeus 1.49.34); idem, fragment 832 (SVF 2:226–227 = Philo *Questions and Answers on Genesis* 1.75); idem, fragment 833 (SVF 2:227, assigned to various passages in Philo); idem, fragment 836 (SVF 2:227 = Aetius *Placita* 4.21); idem, fragment 879 (SVF 2:235–236 = Chalcidius *On the "Timaeus"* 220); Philo *On the Creation of the World* 117; Iamblichus *On the Soul* 12 (trans. Finamore and Dillon 2002:37). For discussions of the Stoic division of the soul, see Safty 2003:293–297, Dillon 1996:102, and Stein 1886–1888:1.119–125.

[43] *Stromateis* 4.23.150.2: ὁ μὲν οὖν ἄνθρωπος ἁπλῶς οὕτως κατ' ἰδέαν πλάσσεται τοῦ συμφυοῦς πνεύματος·

[44] *Stromateis* 4.23.150.2: ὁ δέ τις ἄνθρωπος κατὰ τύπωσιν τὴν ἐγγινομένην τῇ ψυχῇ ὧν ἂν αἱρήσηται χαρακτηρίζεται.

Adam, as regards his formation, was perfect. For he lacked none of the things that characterize (χαρακτηριζόντων) the form and shape of a human being."[45] At the root of both this description and that of the tenth and highest part of the anthropological decalogue is the idea of a person freely choosing the things that characterize or—to capture the overtones of χαρακτήρ—'inscribe' themselves into a person's soul.

Further into his excursus on the Decalogue, Clement returns to this contrast between the lower eight and the upper ninth and tenth human faculties (136.4). The two tablets on which the Ten Commandments were written are said to indicate that "the commandments are given to the two spirits, both the one formed and the ruling faculty (ἡγεμονικῷ)."[46] The difference between the two spirits corresponds to the difference between sense perception and the mental process (137.1). This describes the eighth and the ninth faculties of the anthropological decalogue. The eighth element is conceived to be part of the realm of sense perception. The tenth element is referred to earlier in Book 6 (103.5), where Clement compares the perfected gnostic to Moses, whose face shone (Exodus 34.29). This glorified face is called the "characteristic mark (ἰδίωμα χαρακτηριστικόν) of the just soul"—once again, his preferred terminology for the tenth faculty. For Clement, a person incorporates this faculty into his life as his highest divine power. When the gnostic is perfected as far as his human nature allows, this radiance unites him to God (104.1).

To bring all these threads together: Clement's eighth faculty is the breath or spirit that God breathed into man at his creation (Genesis 2.7). It is common to all people, and it provides for them their structure, both their physical makeup and their essence. This formed spirit is part of the faculties of sense perception, although it is the highest of these, placed above language (i.e. the voice) and the capacity for procreation. The Stoic ruling faculty, the ἡγεμονικόν, the governor of all sense perception, is Clement's ninth faculty. His tenth faculty, the characteristic mark of the Holy Spirit, is the highest divine principle, applied by a person as he chooses the things he wishes to imprint upon his soul. The mark of the Holy Spirit transcends the ruling faculty, since it allows a person to be assimilated to God, as far as human nature allows.

The distinction between the eighth and tenth faculties mirrors one of Clement's better-known theological themes, the distinction between the image and the likeness in God's creation of humanity, according to Genesis 1.26: "And God said, 'Let us make man in our image and in [our] likeness.'" Clement takes

[45] *Stromateis* 4.23.150.3: ᾗ καὶ τὸν Ἀδὰμ τέλειον μὲν ὡς πρὸς τὴν πλάσιν γεγονέναι φαμέν· οὐδὲν γὰρ τῶν χαρακτηριζόντων τὴν ἀνθρώπου ἰδέαν τε καὶ μορφὴν ἐνεδέησεν αὐτῷ.

[46] *Stromateis* 6.16.136.4: εἰκότως τοίνυν αἱ δύο πλάκες τοῖς δισσοῖς πνεύμασι τὰς δεδομένας ἐντολὰς τῷ τε πλασθέντι τῷ τε ἡγεμονικῷ τὰς πρὸ τοῦ νόμου παραδεδομένας ἀλλαχῇ εἴρηνται μηνύειν.

'image' in the biblical text to refer to the state of man at his creation, and 'like-ness' to his eventual acquisition of perfection.[47] Such distinctions between image and likeness run throughout the patristic tradition.[48] It is no different here: the eighth faculty, the formed spirit, corresponds to the image of God; the tenth faculty to the likeness. All people have the eighth; only the redeemed can truly have the tenth.

Clement's anthropological decalogue is intended to argue that the standard Stoic anthropology is incomplete, lacking two prominent aspects of man taught in the Scriptures and recognized by experience. By inserting into the Stoic scheme the faculty of the formed spirit (his new eighth element), Clement teaches that the divine image is common to all people. The Stoics had a notion of human nature's being stamped by the divine, but nothing in their eightfold anthropology made this explicit. Clement's inclusion of the mark of the Holy Spirit as the tenth and highest faculty shows that he regarded the ruling faculty (the ἡγεμονικόν) alone as unable to account for the way people could be assimilated to the likeness of God. There must be a faculty higher than the ἡγεμονικόν. After all, everyone has a ruling faculty, but not everyone who diligently exercises it becomes divine.

His subordination of the spiritual faculty to the ruling faculty, and not vice versa, shows that he held God's spirit to be an essential part of every human being, and not just the elect or the clever. For Clement, the spiritual faculty is more central and basic to human existence than is even governance over the senses. It is possible to detect here a tacit criticism of Valentinianism, which sorted people into three categories—corporeal, soulish, and spiritual—and which taught that only the elect were spiritual. For Clement, the spirit, imparted in creation, is found universally. All people are inherently spiritual, in that they carry the breath of God. The only faculty that might be missing from, or at least minimized in, a person is the tenth. This principle is applied in Book 2, where he urges his readers that they need to take the initiative and overleap the nine faculties to the tenth, the mind.[49] The active presence of the tenth faculty is not guaranteed without effort.

Clement developed his intricate anthropological decalogue to provide a coherent and distinctively Christian account of the world, an account that did not merely repeat the philosophical tradition, but responded with an alternative that was just as, if not more, numerically harmonious. And he ensured that its essence was orthodox, not Valentinian.

---

[47] Clement *The Instructor* 1.11.97.2, 1.12.98.3; idem *Stromateis* 2.22.131.2, 6.

[48] See e.g. Crouzel 1956:67–70; Graef 1952; A. Hamman 1987; Ladner 1953; Merki 1952:44–59.

[49] See p. 130 above. Compare also Irenaeus *Against Heresies* 5.6.1, who like Clement acknowledges that all have the image of God in their formation (*in plasmate*), but not all have the likeness established by the Spirit.

Clement Transfigures Marcus' Transfiguration

In the second half of *On the Decalogue*, Clement interprets the significance of eight of the Ten Commandments (*Stromateis* 6, 137.2–148.4). He had planned to treat his subject "in a cursory manner" (133.1: κατὰ παραδρομήν), that is, to discuss briefly only several of the Ten Commandments. But, absorbed by the topic, he wound up writing more than he anticipated, resulting in lengthy asides and confusion over the numbering of the Commandments.[50] When he reaches the Commandment pertaining to the observation of the Sabbath (137.4), Clement begins a lengthy excursus. He argues that God needs no rest, so the Sabbath rest, really intended for us, indicates cessation from sin, our enlightenment by wisdom and knowledge, and our establishment in dispassion (137.4–138.2). He then says that his discussion (138.5: λόγος) has slipped into the theme of the hebdomad and ogdoad, and here he begins a self-acknowledged tangent (ἐν παρέργῳ), a treatment of the number symbolism of six, seven, and eight. He says, "The ogdoad is likely to be chiefly a hebdomad, and the hebdomad, a hexad, at least apparently. The first is likely to be chiefly the Sabbath, but the hebdomad, a woman worker."[51] This initiates an arithmological excursus in which Clement explains the paradox of how an eight can be considered Sabbath-like, and seven worker-like.

So Clement moves from Sabbath to seven to a larger discussion of the symbolism of six, seven, and eight. The second shift may seem unintuitive, but it fits his overall approach to numbers. Seven symbolizes for Clement a place of rest and completion, and eight a higher state wherein the divine presence resides. Many times, when Clement invokes a symbolic seven, he then notes the need to transcend it to reach the number eight. The hebdomad symbolizes rest, but it is surpassed by the ogdoad, wherein is the promise of gnostic perfection. For example, in Book 6 (108.1), Clement says that those who reach the highest levels of perfection have not remained in the hebdomad of rest, but have advanced into the inheritance of the benefit of the ogdoad (ὀγδοαδικῆς εὐεργεσιάς). Elsewhere Clement says the ark of the covenant symbolizes the ogdoad, and the cherubim symbolize the rest that remains with the glorifying spirits. "Rest," of course, points to the number seven.[52] And in a third example, Clement applies this theme to Ezekiel 44.26–27, where Ezekiel's requisite purification of seven days is said to represent the completion of creation and the ritual observance of rest. The propitiation (ἱλασμόν), which makes acquiring the

---

[50] In my view, this confusion also results from a scribal intrusion. For the full argument see Kalvesmaki 2006:404–411.

[51] κινδυνεύει γὰρ ἡ μὲν ὀγδοὰς ἑβδομὰς εἶναι κυρίως, ἑξὰς δὲ ἡ ἑβδομὰς κατά γε τὸ ἐμφανές, καὶ ἡ μὲν κυρίως εἶναι σάββατον, ἐργάτις δὲ ἡ ἑβδομάς· On ἐργάτις see Proverbs 6.8a LXX.

[52] *Stromateis* 5.6.36.3.

promise possible, is brought on the eighth day (not specified in the Septuagint). According to Clement, Ezekiel's references to seven and eight days point to the doctrine of the hebdomad and ogdoad.[53]

Clement does not think this is a private conceit. Quoting from Clement of Rome, he discusses his namesake's treatment of Psalm 33.13 (34.12): "Who is the man desiring life / yearning to see good days?"[54] Breaking into the quotation, Clement comments, "He"—referring to either Clement of Rome or the psalmist—"then adds the gnostic mystery of the hebdomad and the ogdoad." Then the citation of Clement of Rome and the next two verses of Psalm 33 resumes: "Stop (παῦσον) your tongue from evil / and your lips from uttering deceit. / Turn away from evil and do good, / seek peace and pursue it."[55] So to Clement, either David or Clement of Rome knew of and invoked the symbolism of seven and eight. The mystery of the hebdomad is in the verb παῦσον, translatable as 'rest!' and thereby invoking the seventh day, the Sabbath. The number eight is referred to perhaps in the last verse, in commands to turn away (ἔκκλινον), seek (ζήτησον), and pursue (δίωξον), presumably the realm of the ogdoad.

For Clement, not only the Bible and his predecessors were attuned to the doctrine of the hebdomad and ogdoad. Even Plato knew it. Clement cites the *Republic*: "Now when seven days had reached the [spirits] that were in the meadow, on the eighth they were obliged to proceed on their journey and arrive on the fourth day."[56] He claims that Plato prophesies the Lord's Day. His phrase "seven days" refers to the motions of the seven planets, hastening to their goal of rest; the meadow constitutes the eighth, fixed sphere (τὴν ἀπλανῆ σφαῖραν), and the journey represents the passage beyond the planets to the eighth motion and day. This eighth level, the fixed sphere, Clement elsewhere calls Atlas, the dispassionate pole, and the unmoved aeon.[57] For Clement, the gnostic Christian should be sensitive enough to the doctrine to detect it in certain keywords that would elude the careless or less disciplined reader of Plato, the Scriptures, or the luminaries of the Church.[58]

Variations of Clement's doctrine of the hebdomad and ogdoad were taught by a number of Valentinians. In Irenaeus' extended group of Valentinians, seven is the symbol of the Demiurge, and eight of Akhamoth or Wisdom. Heracleon

---

[53] *Stromateis* 4.25.158–159.

[54] *Stromateis* 4.17.109.1–2, citing Clement of Rome *Letter to the Corinthians* 22.

[55] *Stromateis* 4.17.109.2: εἶτα ἑβδομάδος καὶ ὀγδοάδος μυστήριον γνωστικὸν ἐπιφέρει· "παῦσον τὴν γλῶσσάν σου ἀπὸ κακοῦ καὶ χείλη σου τοῦ μὴ λαλῆσαι δόλον· ἔκκλινον ἀπὸ κακοῦ καὶ ποίησον ἀγαθόν, ζήτησον εἰρήνην καὶ δίωξον αὐτήν."

[56] *Stromateis* 5.14.106.2–4, citing Plato *Republic* 10.616b: ἐπειδὴ δὲ τοῖς ἐν τῷ λειμῶνι ἑκάστοις ἑπτὰ ἡμέραι γένοιντο, ἀναστάντας ἐντεῦθεν δεῖ τῇ ὀγδόῃ πορεύεσθαι καὶ ἀφικνεῖσθαι τεταρταίους.

[57] *Stromateis* 5.6.36.2.

[58] For more examples (such as *Stromateis* 7.10.57.4–5) and analysis of the seven and eight in Clement of Alexandria, see Itter 2009:39–51.

makes six, seven, and eight the symbols of material evil, the aeonic realm, and spiritual perfection respectively.[59] Theodotus, a second-century Valentinian whose doctrines fascinated Clement, held to a similar doctrine.[60]

So in his *On the Decalogue*, after introducing the Sabbath, Clement itemizes the various properties of each of the numbers six, seven, and eight (138.6–140.2).[61] Much of this discussion, and later parts of *On the Decalogue* (141.7–142.1), build upon a Jewish tradition of arithmology, a tradition evident in the writings of Philo and Aristobulus, who argued that Jewish law and custom harmonize with Hellenic philosophy.[62] Clement spends the most time on the number six, pointing out its role in the cosmogony, in the course of the sun, and in the cycles of plant life. He appeals to the importance of six in embryology and to the arithmetical properties that led the Pythagoreans to make it a symbol of mediation and marriage. Six is a function of generation and motion. Seven is depicted as motherless and childless, like its arithmetical properties, since seven neither is the product of, nor produces, any of the numbers in the Decad.[63] Seven was traditionally assigned to Athena, the virgin born without a mother. Clement, however, takes the Pythagorean epithet to refer allegorically to the Sabbath and the form of rest in which "there is neither marrying nor being married" (Matthew 22.30). The ogdoad is briefly described as the cube, the fixed sphere, and a participant in the Great Year.[64]

---

[59] Heracleon, frags. 15, 18, 40 (= Origen *Commentary on John* 10.248–250, 13.69–72, 13.416–426). Wucherpfennig denies that fragment 40 refers to the doctrine of the six, seven, and eight, and suggests rather that it refers to the seventh day of Creation and God's restoring human nature to its original good standing (2002:320–321). But Heracleon (at Origen *Commentary on John* 13.424) discusses a nature that is "depicted" (χαρακτηρίζεται), not "restored." In this passage natures are not transformed (as Wucherpfennig's reading would require), they are revealed.

[60] Clement of Alexandria *Epitomes* 3.63. On Theodotus, see DECL 571, Thomassen 2006:28, and Kalvesmaki 2008.

[61] Clement's excursus (§§138.5–145.7) has been studied most thoroughly and skillfully by Delatte (1915:229–245), who was also the first to recognize that Clement's exegesis of the Transfiguration depends upon the teachings of Marcus. Delatte identifies Clement's sources for his arithmology (1915:234–235). See also parallels in the *Theology of Arithmetic* at sections devoted to 6, 7, and 8.

[62] See Delatte 1915:233, for the scope and evidence of Clement's direct or (more likely) indirect use of Aristobulus. Although Clement uses Aristobulus, this is probably mediated through someone like Hermippus of Berytus, a grammarian (possibly Jewish or Christian) of the early second century, whose lost treatise *On the Hebdomad* is mentioned by Clement at §145.2. Compare also *Stromateis* 5.14.107.1–108.1, Clement's catena of quotations from classical authors who praise the number seven.

[63] There are many ancient parallels. See e.g. Aristotle, fragment 203 (= Alexander of Aphrodisias *Commentary on Aristotle's "Metaphysics"* 38), Philo *On the Creation of the World* 100, *Theology of Arithmetic* 41.30, Theon of Smyrna *Mathematics Useful for Reading Plato* 103.14–16, Nicomachus of Gerasa in Photius *Biblioteca* 144b.

[64] Fixed sphere: τὴν ἀπλανῆ ... σφαῖραν: see *Stromateis* 5.106, cited above. Because of precession, the drifting of the earth's axis across the stars, the zodiac appears over the centuries to rotate slowly around the earth. The time it takes for one house of the zodiac to return to its starting point,

Up to this point, Clement has marked only the differences between six, seven, and eight. There is nothing yet to justify his claim at 138.5 that there is an identity between them, or a transformation from one to another. To make this connection Clement turns to the Transfiguration (140.3). He locates the event on the eighth day, and the Lord as the fourth person (after Peter, James, and John). The Lord ascends the mountain, and at the appearance of Moses and Elijah (i.e. two more persons) becomes the sixth. Clement reckons the voice of God as the seventh character, and Jesus is made manifest as the eighth, God.

This passage owes much to Marcus' explanation of the Transfiguration (see chapter 4). As Niclas Förster has suggested, Clement's knowledge of Marcus relied exclusively upon Irenaeus.[65] Clement does not employ Marcus' doctrine anywhere else, and his departures from Irenaeus' account are in line with his other adaptations of *Against Heresies*.[66] Sometimes he quotes Irenaeus verbatim; other times he adds to the account, so as to make Marcus' interpretation his own (underlining indicates textual parallels):

ταύτῃ τοι ὁ κύριος τέταρτος ἀναβὰς εἰς τὸ ὄρος ἕκτος γίνεται καὶ φωτὶ περιλάμπεται πνευματικῷ, τὴν δύναμιν τὴν ἀπ' αὐτοῦ παραγυμνώσας εἰς ὅσον οἷόν τε ἦν ἰδεῖν τοῖς ὁρᾶν ἐκλεγεῖσι, δι' ἑβδόμης ἀνακηρυσσόμενος (15) τῆς φωνῆς υἱὸς εἶναι θεοῦ, ἵνα δὴ οἳ μὲν ἀναπαύσωνται πεισθέντες περὶ αὐτοῦ, ὁ[67] δέ, διὰ γενέσεως, ἣν ἐδήλωσεν ἡ ἑξάς, ἐπίσημος[68] ὀγδοὰς ὑπάρχων φανῇ, θεὸς ἐν σαρκίῳ τὴν δύναμιν ἐνδεικνύμενος, ἀριθμούμενος μὲν ὡς ἄνθρωπος, κρυπτόμενος δὲ ὃς ἦν·

So on this [eighth day][69] the Lord, as the fourth, after ascending the mountain becomes a sixth and is radiated by a spiritual light, laying bare his power—as far as it is possible for those chosen to see to

---

calculated by modern astronomers at 26,000 years, is called the Great Year, the length of which was a topic in classical antiquity. See e.g. Plato *Republic* 8.546.

[65] Förster 1999:252n207.

[66] See Behr 2000, le Boulluec 1982:707–713, and Patterson 1997. For a well-known parallel see *Stromateis* 7.18.109.2–110.1, adapting Irenaeus *Against Heresies* 5.8.3.53–84, and analysis in Hort and Mayor 1902 and Patterson 1997.

[67] Substituting Delatte's ὁ for the ὃ in SC and the οἱ (dittography from the previous line) in the manuscript.

[68] I delete the comma here, following Sagnard 1947:378 and Dupont-Sommer 1946:47. On the coined phrase 'episēmos ogdoad,' see p. 72 above.

[69] The precise referent of ταύτῃ is missing from the ANF and SC translations. It must refer to the ogdoad because that was the topic of the previous section and because ταύτῃ corresponds to Marcus' μετὰ τὰς ἓξ ἡμέρας. It may possibly refer to the last word of §140.2, ἀνταποδόσεως, but even this term Clement gives the nuance of eightness. My reconstruction makes sense in light of Luke 9.28, which places the Transfiguration on the eighth day, unlike Matthew 17.1 and Mark 9.2, both of which place it "after six days." Clement follows Luke; Marcus opts for Matthew and Mark.

perceive—and heralded by the seventh, the voice, to be the Son of God, so that those who are persuaded about him might rest, while he, being an episēmos ogdoad, might be manifest through his generation (which the hexad makes clear) as God, demonstrating his power in a bit of flesh: numbered as man, but keeping hidden who he was.

(Clement of Alexandria *Stromateis* 6.140.3.10)

Τούτου τοῦ λόγου καὶ τῆς οἰκονομίας ταύτης καρπόν φησιν ἐν ὁμοι-
ώματι εἰκόνος πεφηνέναι ἐκεῖ(νον) τὸν μετὰ τὰς ἓξ ἡμέρας τέταρτον
ἀναβάντα εἰς τὸ ὄρος καὶ γενόμενον ἕκτον, τὸν καταβάντα καὶ κρατηθέντα ἐν τῇ Ἑβδομάδι, ἐπίσημον Ὀγδοάδα ὑπάρχοντα καὶ ἔχοντα ἐν
ἑαυτῷ τὸν ἅπαντα τῶν στοιχείων ἀριθμόν.

[Marcus] says that the fruit of this account and this plan is that he was manifest in the likeness of an image [Romans 1.23], he who, after six days, ascended the mountain as the fourth and became a sixth, he who descended and was held in the Hebdomad, being an episēmos ogdoad and possessing within himself every number of the oral letters.

(Irenaeus *Against Heresies* 1.14.6.272–277)

Armand Delatte claims that the two passages complement and explain each other. Clement's description of Jesus as ἐπίσημος is intelligible only when we consider that Marcus calls Jesus this because his name consists of six letters.[70] Likewise, Marcus' peculiar phrase, that Christ was held in the hebdomad, is rendered intelligible by Clement's version, where the seventh is identified with the voice that possesses Jesus at his baptism and declares him to be the Son of God.[71] According to Delatte, both Clement and Marcus represent Christ as six, seven, and eight inclusively, numbers that symbolize the Incarnation. Thus Clement's version of Marcus' Transfiguration account synthesizes the previous sections discussing the properties of six, seven, and eight (138.5–140.2), and help demonstrate Clement's claim, since Christ himself is six, seven, and eight.

The problem with Delatte's overall interpretation is that strictly speaking, Marcus and Clement identify Christ not with seven but only with six and eight. Seven is reserved in Clement for the voice of God, not Jesus. Marcus regards the hebdomad as the seven vowels, which receive an eighth letter so as to equal the size of the other two levels of the aeonic alphabet. Thus six, seven, and eight are not indiscriminately all symbols of Jesus. And what started Clement on

---

[70] Delatte 1915:238. Irenaeus *Against Heresies* 1.14.4.
[71] Delatte 1915:238–239.

this tangent in the first place was the proposal that the ogdoad is a hebdomad, and the hebdomad a hexad (138.5). This requires a transitive relationship, of 6 → 7 → 8, or 8 → 7 → 6. But neither Clement's nor Marcus' account of the Transfiguration suggests that Jesus went from being the sixth to becoming the eighth, via the seventh. Clement may have originally (138.5) proposed to demonstrate a transformation of six into eight, but that does not occur here (at 140.3). The only numerical transformation in this passage is from four to six, when Jesus becomes the sixth after the appearance of Moses and Elisha. Clement states (as does Marcus) that Jesus is *simultaneously* six and eight—the 'episēmos ogdoad'—without implying any transition between the two (at least, not in 140.3). Both passages emphasize the differences, not the transformations, among six, seven, and eight. Clement's version constitutes a Christian counterpart to the "secular" arithmology of 138.6–140.2, which lists common ideas in the Greco-Roman world about six, seven, and eight. Clement uses a single event, the Transfiguration, to depict the Christian understanding of the symbolic significance of six (generation), seven (the voice), and eight (divinity), arraying the symbolism for all three numbers in a single, static image.

A closer look at the vocabulary of 140.3 bears this out. I have mentioned above the care Clement shows when he reads other authors and "discovers" in them the doctrine of the hebdomad and ogdoad. Sometimes this is made explicit by keywords, such as those in Plato's *Republic*. Other times, the association emerges through ideas, not terms, as in Clement of Rome and Psalm 33 (34). Clement considers Scripture to have been written with extreme care. There are no superfluous words, and each word is chosen for its symbolic overtones, no matter how subtle. He composes his explanation of the Transfiguration with the same care. His words are tinged with overtones of number symbolism, and thereby present a complex Christian arithmology of six, seven, and eight.

Clement gives to Christ the epithet 'episēmos ogdoad.' This epithet, which features prominently in Marcus' theology, conjoins six and eight, a mathematical and theological paradox. Six represents the created, material world; and eight, spiritual perfection, the divine realm. Notice how the paradox is reflected at the end of 140.3, where Jesus is "numbered as man," but hidden as "he was." Six corresponds to "man," and eight, to his constant state of being, i.e. God. Between these two phrases, however, is θεὸς ἐν σαρκίῳ τὴν δύναμιν ἐνδεικνύμενος, a phrase that can also be interpreted as a cipher for 'six-eight': σαρκίῳ is six; θεός and δύναμις eight. The entire clause, from ὁ δέ to the end, reiterates in three compact phrases the mystery of the Incarnation as a combination of six and eight.

Likewise, the first instance of δύναμις (line 13), just as the second (line 18), should also suggest the number eight. This is consistent with Clement's exposition, since this first instance describes how Christ revealed his divinity, as much as his companions could manage. Divinity is often represented as the number eight in Clement. The spiritual light in which Jesus is cloaked is the radiance of this "eightness." This entire phrase alludes to the eightness of the Transfiguration, and complements the next phrase, which explicitly identifies seven with the voice that permits his disciples to find rest: ἑβδόμης, φωνῆς, and ἀναπαύσωνται all cross-resonate. By finding rest, the disciples, who are products of generation and therefore symbolized by six, move from the realm of six into seven.

Read this way, most of the text in 140.3 that is not emphasized (from καὶ φωτὶ το περὶ αὐτοῦ), which has no parallel in Marcus or Irenaeus, constitutes a miniature Christian arithmology on eight, then seven. It moves on to six—the generative aspect of Jesus—where Clement picks up again from Irenaeus' text, and then ends in a terse meditation on the Incarnation as a combination of six and eight (from Θεὸς ἐν σαρκίω to ὃς ἦν). Thus the material not in Marcus/Irenaeus is Clement's careful augmentation.

From here, Clement's account of the Transfiguration moves not to the baptism of Jesus, as Marcus' does, but to the order of the alphabet and numerical notation. This too is a theme for Marcus, but Clement pursues the matter further. He begins by explaining that six is included in the order of the numbers, but that the sequence of the alphabet shows that the ἐπίσημον is not written with a letter. That is, numerals, using the alphabetic system of numeration, follow the sequence α΄, β΄, γ΄, δ΄, ε΄, ϛ΄, ζ΄, η΄, and so on. The number six is represented by the ἐπίσημον. But when the alphabet is written out—α, β, γ, δ, ε, ζ, η, and so on—the ἐπίσημον is not written. He explains that the difference between the two sequences is created by the intrusion of the ἐπίσημον, which disrupts the alphabet, a disruption that he takes as a cipher for his doctrine of the six and seven, and subsequently of the seven and eight. Here is the relevant text:

τῇ μὲν γὰρ τάξει τῶν ἀριθμῶν συγκαταλέγεται καὶ ὁ ἕξ, ἡ δὲ τῶν στοιχείων ἀκολουθία ἐπίσημον γνωρίζει τὸ μὴ γραφόμενον. ἐνταῦθα κατὰ μὲν τοὺς ἀριθμοὺς αὐτοὺς σώζεται τῇ τάξει ἑκάστη μονὰς εἰς ἑβδομάδα τε καὶ ὀγδοάδα, κατὰ δὲ τὸν τῶν στοιχείων ἀριθμὸν ἕκτον γίνεται τὸ ζῆτα, καὶ ἕβδομον τὸ η¯. [2] Ἐκκλαπέντος[72] δ᾽ οὐκ οἶδ᾽ ὅπως τοῦ ἐπισήμου εἰς τὴν γραφήν, ἐὰν οὕτως ἐπώμεθα, ἕκτη μὲν γίνεται ἡ ἑβδομάς, ἑβδόμη δὲ ἡ ὀγδοάς·.

For the [number] six is included in the order of the numbers, but the sequence of the oral letters makes known that the ἐπίσημον is

---

[72] Ms reading. Stählin 1909: Εἰσκλαπέντος.

unwritten. Thus, according to the numbers themselves, each monad is preserved in sequence, up to the hebdomad and the ogdoad. But according to the number of oral letters, the *zeta* becomes sixth, and the *eta* seventh. But when the ἐπίσημον—I don't know how—slips into[73] writing (should we pursue it in this manner) the hebdomad becomes the sixth [letter], and the ogdoad the seventh.[74]

(Clement of Alexandria *Stromateis* 6.140.4–6.141.1)

To unravel this cryptic passage, preliminary comments on two aspects of Greek grammar are in order. First, as already discussed, the grammarians distinguished στοιχεῖον, the oral letter, from γράμμα, the written.[75] Clement also holds to this distinction. The phrases τὸ μὴ γραφόμενον and εἰς τὴν γραφήν show that he sees the ἐπίσημον as dwelling in the written sphere, not the oral. That is, the ἐπίσημον is seen, not heard.

Second, Clement is not discussing the *digamma*, the archaic Greek letter derived from the Phoenician *waw*. His comments here are frequently misread because modern readers conflate the ἐπίσημον and the *digamma*. It is common knowledge today that the earliest Greek alphabets included in the sixth place the Phoenician letter *waw*, first written like a Y, but later as ϝ.[76] The *digamma* dropped out of use in the Greek language, but its written representation was preserved in the Milesian system of numeration. The ἐπίσημον is seen as the direct descendant of the obsolete *waw*. But late antique and medieval treatments of the *digamma* show no awareness that it was the ancestor of the numeral six. Greek grammarians in late antiquity did not even assign the *digamma* a place in the sequence of the alphabet. Further, no ancient discussions of the παράσημα— the nonalphabetic numerals—mention the letter *digamma*.[77] An ancient scho-

---

[73] Or "slips out of." See discussion below.

[74] That is, the number seven comes to be represented by the sixth oral letter.

[75] See p. 63 and n4 above.

[76] The term *digamma*, attested in the postclassical period, comes from the character's looking like one uncial gamma superimposed on another. See LSJ 752a s.v. "ϝ," and Larfeld 1898:294.

[77] All ancient references known to me concerning the ἐπίσημον are discussed above, pp. 66–88. *Scholia on Dionysius Thrax* 1.3:496.6–7 appears at first glance to identify the *digamma* with the Greek numeral for six. The scholiast entertains the question, Why are there twenty-four letters (γράμματα) when there are other characters and inscribed figures, and other nations have their own letters, and there are certain other figures: "the digamma, the koppa, the so-called παρακύϊσμα, the insignia, and things written alongside letters, and the crown?" (Διὰ τί δὲ κδ´ ἔφη εἶναι τὰ γράμματά εἰ γὰρ γράμματά εἰσιν οἱ χαρακτῆρες καὶ οἱ ξυσμοί, γράμματα δὲ καὶ τὰ παρὰ Χαλδαίοις καὶ Αἰγυπτίοις, καί τινα ἔτερα, τὸ δίγαμμα καὶ τὸ κόππα καὶ τὸ καλούμενον παρακύϊσμα, καὶ τὰ σημεῖα, καὶ τὰ παρεγγραφόμενα τοῖς στοιχείοις, καὶ ἡ κορωνίς, καὶ εἴ τι τοιοῦτον, ἀτόπως φησὶν ὅτι κδ´ ἐστίν.) This passage has guided scholars, including LSJ 1562a, s.v. "Μ," and LSJ, Supplement 114, s.v. "ˣπαρακύϊσμα," to define παρακύϊσμα—a hapax legomenon—as the term for the numeral ϡ. But this presumes that the author identified the *digamma*

lium on Dionysius Thrax precludes such an association. This scholiast entertains the theoretical objection that, because the *digamma* is a letter, Dionysius Thrax's claim that there are twenty-four written letters must be faulty. The objection runs: both a character (χαρακτήρ) and a name (ὄνομα) are concomitant with every oral letter; the *digamma* has both, so it too should be reckoned with the oral letters. The scholiast lays out several responses to this argument, one of which runs: "Again, every character (χαρακτήρ) of the oral letters designates a number. For the α indicates the number one, and the β, two, and so forth. So therefore, if the character of ϝ doesn't indicate a number, it is clear that it is not an oral letter."[78] So this scholiast regarded the *digamma* as having no corresponding numeral and therefore no place in the order of the alphabet. The other parallel scholia discussing the *digamma* also do not associate it with the number six or any ordinal place in the alphabet.[79]

This explains why Ptolemy, in his *Harmonics*, uses both the *digamma* and the ἐπίσημον in the same sentence to refer to two different things: the *digamma* to a musical tone and the ἐπίσημον to a numeral.[80] In a sixth-century Greek text, *The Mystery of the Letters*, the godless Greeks are accused of moving the *waw* from its proper place and placing it after the *nu*. God is said to have providentially used the disruption to make the *waw* a symbol of Christ.[81] To this author, the Phoenician *waw* became not the *digamma* but the *omicron*!

All these late antique texts show that any original association there may have been between the numeral six and the *digamma* had been lost by the second century. This helps clear up a vexing textual problem at 140.4–141.2, one that

---

with the numeral ϛ and intended to list all three nonalphabetic numerals. This should be shown, not assumed. In this passage παρακύϊσμα can be read with δίγαμμα and κόππα as a threesome, but it might be better grouped with τὰ σημεῖα to form a second pair of terms. Given the root meaning of παρακύϊσμα—a κύημα is a fetus—it is very difficult to see how the character ⟩ could be inferred. Jannaris 1907:39, suggests that the ⟩ "is a παρακλῖνον γέννημα," "a slanting letter," but offers no explanation of how it resembles "offspring." In reality, we have no clue what παρακύϊσμα means. This scholium on Dionysius Thrax is the only ancient text that might possibly be interpreted to connect the *digamma* with the numeral six, a shaky foundation given other arguments below.

[78] *Scholia on Dionysius Thrax* 1.3:187.22–25. Ἔτι πᾶς χαρακτὴρ στοιχείων σημαίνει ἀριθμόν· καὶ γὰρ τὸ α σημαίνει τὸν ἕνα ἀριθμόν, καὶ τὸ β τὸν δύο, καὶ ἑξῆς· εἰ ἄρα οὖν ὁ χαρακτὴρ τοῦ ϝ οὐ σημαίνει ἀριθμόν, δῆλον ὅτι οὐκ ἔστι στοιχεῖον.

[79] *Scholia on Dionysius Thrax* 1.3:34.15–23; 2.1:76.32–77.12. At first glance, the Georgian alphabet seems to provide evidence that late antique grammarians knew about the connection between ϝ and ϛ. The fifth, sixth, and seventh letters are ე [e], ვ [v], and ზ [z], and the alphabet was used for numerals in the fashion of Greek. But Mouraviev has demonstrated that the placement of extra Georgian letters, such as ვ in the alphabet had nothing to do with the Semitic alphabets, but was the careful, deliberate work of a phonologist (1984). The phonetic equivalent of *waw* is the 22nd letter, ჳ [ü/w], assigned the value of 400 in Georgian alphabetic numeration.

[80] Ptolemy *Harmonics* 2.1.

[81] *The Mystery of the Letters* 31–33 (Bandt 2007:170–174; cf. 227).

is critical to understanding the entire passage. It is unclear whether Clement thought the letter slipped into writing, or fell out of it, since the prepositions in ἐκλαπέντος and εἰς τὴν γραφήν are contradictory. Some scholars have argued that the text should read εἰσκλαπέντος (which would have Clement regard the ἐπίσημον as entering the alphabet), others as ἐκ τῆς γραφῆς (to have him see the character fall into disuse).[82] The former are correct, but they are seemingly unaware of the grammatical background, just discussed.[83] Delatte's proposal, which depends upon the latter group, has Clement, in agreement with modern scholarship, meditating on the development of the Greek alphabet from its Phoenician roots. But this is not Clement's point. He was interested primarily in the difference between the alphabet and Greek numeration. Like others in his day, Clement saw no connection between the *waw* and the numeral six, since he considered the latter as a purely written symbol, not a spoken one. Thus the text should read Εἰσκλαπέντος δ᾽ οὐκ οἶδ᾽ ὅπως τοῦ ἐπισήμου εἰς τὴν γραφήν.[84] The numeral six, the ἐπίσημον, somehow entered *into* the writing system—Clement admits his ignorance on the historical specifics—and thus disrupted the order of the alphabet.

This makes Clement's allegory clearer: the ἐπίσημον symbolizes Christ, who enters the writing of the world and alters the constitution of its oral letters/ elements (στοιχεῖα). Clement plays on the ambiguity of στοιχεῖον, treating it primarily as a letter of the alphabet, but also alluding to its alternate meaning as an element of the universe. He regards the inconcinnity between the alphabet and the numbering system to be the key to interpreting the effect of the Incarnation on creation. This same inconcinnity explains the numbers latent in the Transfiguration. There on the mountain, Jesus is revealed as the 'episēmos ogdoad,' the number eight in the guise of the numeral ς. The number eight is the unknowable God, the ς is his entry into the writing system. Only here do we encounter the transition 6 → 7 → 8, and it pertains to the movement

---

[82] Delatte 1915:241.

[83] Delatte (agreeing with Serruys) rejected Stählin's argument (agreeing with Lowth) for the first option, since it rested on the view that the last sentence in Clement's paragraph purports to say that the number seven then took the sixth place, and the number eight the seventh. Stählin argued for the first option by suggesting that the numbers themselves move. Of course, they do not. By affirming the second option, that the ἐπίσημον fell out of writing, Delatte influenced the SC edition (Descourtieux 1999:342–343), which departs from Stählin's text—otherwise the preferred edition—in favor of the manuscript Laur. V 3.

[84] There are other reasons for accepting the emendation. First, the alternate proposal, Ἐκκλαπέντος δ᾽ οὐκ οἶδ᾽ ὅπως τοῦ ἐπισήμου ἐκ τῆς γραφῆς, requires an alteration of three words, rather than just one. Second, the last sentence in Clement's paragraph does not suggest what Stählin said and what Delatte discounted (see n. 83 above); my translation and explanation clarify Clement's meaning. Third, the alternate proposal suggests that the ἐπίσημον was originally in the alphabet, then disappeared, contradicting Clement's previous sentence, which states that before whatever happened to the ἐπίσημον, *zeta* was sixth and *eta* seventh in the order of the alphabet.

of believers who transcend their humanity. The intrusion of the ς causes the sixth element (στοιχεῖον) to access the seventh, and the seventh element (στοιχεῖον) to access the eighth. Thus the apostles, by trusting in him on the mount of Transfiguration, entered into the rest of the seventh. We shall see below that Clement, by analogy, has the faithful move from the seventh to the eighth, but his interpretation of the Transfiguration (140.3) stops short of this. Instead, Clement now turns to Marcus' teaching on the number six (141.3–7). He draws from parts of Scripture that speak to the doctrine of the ἐπίσημον, and then selects examples from geometry to establish the point he set out to make initially, that the ogdoad is likely to be a hebdomad, and the hebdomad a hexad. His argument runs:

> διὸ καὶ ἐν τῇ ἕκτῃ ὁ ἄνθρωπος λέγεται πεποιῆσθαι ὁ τῷ ἐπισήμῳ πιστὸς γενόμενος ὡς εὐθέως κυριακῆς κληρονομίας ἀνάπαυσιν ἀπολαβεῖν. τοιοῦτόν τι καὶ ἡ ἕκτη ὥρα τῆς σωτηρίου οἰκονομίας ἐμφαίνει, καθ᾽ ἣν ἐτελειώθη ὁ ἄνθρωπος. ναὶ μὴν τῶν μὲν ὀκτὼ αἱ μεσότητες γίνονται ἑπτά, τῶν δὲ ἑπτὰ φαίνονται εἶναι τὰ διαστήματα ἕξ. ἄλλος γὰρ ἐκεῖνος λόγος, ἐπὰν ἑβδομὰς δοξάζῃ τὴν ὀγδοάδα καὶ "οἱ οὐρανοὶ τοῖς οὐρανοῖς διηγοῦνται δόξαν θεοῦ." οἱ τούτων αἰσθητοὶ τύποι τὰ παρ᾽ ἡμῖν φωνήεντα στοιχεῖα. οὕτως καὶ αὐτὸς εἴρηται ὁ κύριος "ἄλφα καὶ ὦ, ἀρχὴ καὶ τέλος," "δι᾽ οὗ τὰ πάντα ἐγένετο καὶ χωρὶς αὐτοῦ ἐγένετο οὐδὲ ἕν."

So also, it is said that in the sixth [day] the human was made, becoming faithful to the ἐπίσημον, so as to receive straightaway the rest of the Lord's inheritance. Even the sixth hour of the divine plan of salvation indicates this sort of thing; in it the human was perfected. Indeed, there are seven intermediates of eight things, and there seem to be six intervals of seven things. For there is that other saying, when the hebdomad glorifies the ogdoad and "the heavens declare to the heavens the glory of God." [Psalms 18.2] The oral letters that are our vowels are perceptible types of these things. So also the Lord himself is said to be "alpha and o[mega], beginning and end" [Revelation 21.6], "through whom everything came into being, and without him not even one thing came into being" [John 1.3].

(Clement of Alexandria *Stromateis* 6.141.3–7)

The parallel from Marcus runs:

> Καὶ διὰ τοῦτο Μωϋσέα ἐν τῇ ἕκτῃ ἡμέρᾳ εἰρηκέναι τὸν ἄνθρωπον γεγονέναι· καὶ τὴν οἰκονομίαν δὲ ἐν τῇ ἕκτῃ τῶν ἡμερῶν, ἥτις ἐστὶν ἡ

παρασκευή, <ἐν> ᾗ τὸν ἔσχατον <u>ἄνθρωπον</u> εἰς ἀναγέννησιν τοῦ πρώτου <u>ἀνθρώπου πεφηνέναι, ἧς οἰκονομίας ἀρχὴν καὶ τέλος</u> τὴν <u>ἕκτην ὥραν</u> εἶναι, ἐν ᾗ προσηλώθη τῷ ξύλῳ.

And because of this, Moses said that the human being came into existence on the sixth day, and the divine dispensation, on the sixth day [of the week], i.e. the Day of Preparation, in which the last human being is manifest for the rebirth of the first man. The beginning and end of this divine dispensation was the sixth hour, when he was nailed to the wood.

(Irenaeus *Against Heresies* 1.14.6.280–285)

Καθὼς οὖν αἱ ἑπτά, φησίν, δυνάμεις <u>δοξάζουσι</u> τὸν Λόγον, οὕτως καὶ ἡ ψυχὴ ἐν τοῖς βρέφεσι κλαίουσα καὶ θρηνοῦσα Μάρκον <u>δοξάζει</u> αὐτόν. διὰ τοῦτο δέ καὶ τὸν Δαυὶδ εἰρηκέναι· "Ἐκ στόματος νηπίων καὶ θηλαζόντων κατηρτίσω αἶνον," καὶ πάλιν· "<u>οἱ οὐρανοὶ διηγοῦνται δόξαν θεοῦ</u>."

He [Marcus] says: Therefore, just as the seven powers glorify the Logos, so also the soul in infants, crying and wailing, glorifies Marcus himself. Because of this, David also said, "From the mouth of infants and sucklings, you have perfected praise" [Psalms 8.3 (8.2)], and also, "The heavens declare the glory of God" [Psalms 18.2 (19.1)].

(Irenaeus *Against Heresies* 1.14.8.320–325)

Clement's version is an orthodox, ecclesiastical revision of Marcus' teaching. He notes, in Marcus' words, that the human was created on the sixth day. He omits any mention of Moses, and thereby identifies the sixth day of creation with the day of Christ's crucifixion. Clement parses the phrase "in the sixth [day] the human." Using the same order of cases—dative, then nominative—Clement explains what 'sixth day' and 'human' mean. The sixth day of creation/redemption is the ἐπίσημον, and in that day man becomes faithful to Christ. Clement, again departing from Marcus, says that the purpose of the Creation and Redemption was to have humanity straightaway enjoy the rest of the Lord's inheritance. His wording is precise. In a single phrase he uses ciphers of both seven (ἀνάπαυσιν) and eight (κυριακῆς κληρονομίας).[85] Thus Clement restates that the goal of humanity is to move from the sixth day of Creation, through the Sabbath rest, into the eighth day. He considers the connection between these

---

[85] For parallels see *Stromateis* 5.14.106.2, 6.14.108.1, 7.12.76.4 and *Epitomes* 3.63.

days of creation as tight as the geometric relationship between points and the intervals between them (141.5).

As a further illustration, Clement appeals to Psalm 18 (19), which he emends so that the heavens declare the glory of God *to the heavens* (not in the Septuagint), just as the hebdomad glorifies the ogdoad. The image evoked here is that of the seven planets glorifying the fixed sphere, the same image he uses to interpret Plato's *Republic*, discussed above. In ancient number symbolism the seven planets were closely associated with the seven vowels. So the Lord, who is called *alpha* and *ōmega*, is symbolized in the Psalms by the heavens. The Lord, the creator of all things, is the beginning and the end of all seven vowels. The thrust of 141.6–7 is that Christ constitutes the harmony of the spheres, the one who communicates to all the glory of God.

In this passage Marcus' numbers are more static than Clement's. In the first paragraph, Marcus is concerned with the number six and with showing the relationships among the sixth day of Creation, the crucifixion on the sixth day of the week, and the nailing of Jesus at the sixth hour. He claims the sixth hour was the beginning and the end of redemption, a notion that harmonizes well with the Pythagorean idea of the perfection of the number six.[86] Six does not become anything. The second paragraph, which concerns itself with the number seven, is static. Clement adds the motif of numbers changing and turning into each other, in imitation of the divine dispensation and the Incarnation. He spins these two unrelated passages by Marcus into a new narrative, an orthodox vision of God's becoming man so that man might attain divine unity.[87] The numbers in Clement's new allegory symbolize the vertical transition of the faithful, as they ascend from the material world to the spiritual.

Having read *Against Heresies*, Clement would have known Irenaeus' saucy rhetoric and his favored argument, the *reductio ad absurdum*. Yet Clement seems to take Marcus' exegesis seriously. There is no express sarcasm or criticism, no attempt to show the arbitrariness of his opponent's methods or conclusions. Throughout the *Stromateis* Clement uses the term γνῶσις 'knowledge', to reclaim it from the heretics, the self-declared spiritual, on behalf of his own "ecclesiastics." In like manner he robs Marcus of the symbol 'episēmos ogdoad,' to make of it a sign of Jesus' Incarnation, not of his emanation from and return to the Ogdoad. Both Marcus and Clement consider Jesus to be "noteworthy" because of his association with six. But for Marcus, the sixness is found most immediately in the number of letters in Jesus' name, the number required to generate, with the Tetrad, the 24 letters of the alphabet needed to achieve the aeonic Triacontad.

---

[86] See p. 54n78.
[87] See p. 127 above.

For Clement, the sixness lies not in letter counts but in its symbolism of the human nature of Christ, of the rupture in human discourse that brought about salvation. He ignores any sense of 30, 24, 801, or other numbers that appeal to Marcus. Marcus focuses on the connection between the aeons and the alphabet; Clement, on that between the Incarnation and redemption.

Clement shares with the Valentinians, Monoïmus, and deutero-Simon a fascination with arranging Scripture into arithmetically harmonious structures. Just as they do, he brings to the text a well developed sense of number symbolism. He massages the Scriptures, peers behind individual words, and chases down their overtones, to show how the Bible reveals those structures. The technique works outside the Bible, too. Clement reads ecclesiastical and philosophical literature with an eye to hidden number symbolism. The finest example is his investigation of Stoic anthropology, which he transforms into a Christian one by supplementing the missing parts and molding the structure into a pattern that better fits Scripture. The tactic resembles those of his theological opponents. For Clement, this is no inconsistency, since their error comes from their conclusions, not their tactics.

He does not adhere to all of Irenaeus' four principles for the correct theological use of numbers. Like the Valentinians, Clement draws from human conventions in grammar and numeration to illustrate his theology (contra Irenaeus' second principle). He also quite openly takes preconceived number symbols into the Scriptures and the ecclesiastical tradition, rearranging a bit of the furniture along the way (contra Irenaeus' fourth principle). But other aspects of Clement's number symbolism match Irenaeus'. He has no mathematical arrangement of the godhead, and the symbolism he draws from numbers found in the natural world is based safely on the science of his day (Irenaeus' first and third principles). If Irenaeus were to have any problem with Clement's number symbolism, it would probably revolve around exegesis. But we have already noted how Irenaeus bent his own second and third principles. So if Irenaeus held that Clement professed the apostolic rule of faith—and we have no reason to doubt this—then it is quite probable that Irenaeus would have shown him the same leniency he shows himself. For his part, Clement does not directly criticize Irenaeus. Whatever criticism can be detected is tacit. He faces the same opponents, but does not demand of them standards he fails to attain.

The two different models furnished by Clement and Irenaeus show that in practice, the orthodox theology of arithmetic consisted primarily of a few simple principles. God, in his simplicity, transcends the realm of numbers, which he created. Father, Son, and Spirit are three, not because three is perfect. Rather, three is perfect because Father, Son, and Spirit are the one God. In the rule of truth there reside many numerical symbols that reveal God's ways. To indulge in these and to draw upon number symbolism from culture, science, and mathematics is quite permissible, provided it does not undermine the apostolic faith shared by the churches throughout the world. Such principles were less strictures than signposts, warning the faithful away from the precipices of private fantasy—in a word, heresy.

8

How the Early Christian Theology of Arithmetic Shaped Neo-Platonism and Late Antique Christianity

AFTER THE EARLY THIRD CENTURY, the controversy over the theology of arithmetic disappeared from the Church. The dispute need not have died down. Gnosticizing writings well into the fourth century show a continued interest in speculative number symbolism. But the only orthodox Christian responses to be found come from recherché apologies like Epiphanius' *Panarion*, indebted largely to Irenaeus for its section on Valentinianism.

On the other hand, from the late second through the early fourth centuries number symbolism took on increased importance and new roles in Platonism. The Pythagorean metaphysical innovations begun in the first centuries CE and BCE entered into the Platonist tradition, and in a sequence that echoed the Christian experience: speculative number symbolism, controversy, an attempt to articulate principles, and out of that a sense of what was acceptable in the Platonist tradition.

In this chapter I provide two accounts of the afterlife of the early Christian debate on the theology of arithmetic. The first account treats the role of number symbolism in the metaphysics and exegesis of Platonist philosophy in the late third and fourth centuries. Continuing the story from chapter 2, I continue with the legacy of Plotinus, focusing on an unusual exchange between Theodore of Asine and Iamblichus. I argue that Iamblichus, an important early fourth-century philosopher, played for Platonist orthodoxy a role comparable to that of Irenaeus and Clement. The second account pertains to the orthodox Christian theology of arithmetic. I briefly summarize how the parameters and paradigms established by Irenaeus and Clement shaped later Christian and Byzantine usage.

Platonist Number Symbolism
in the Third and Fourth Centuries

As pointed out in chapter 2, the new philosophical currents that emerged from the first century BCE to the first century CE sought the origins of the world in a single ἀρχή (or a "plurality-in-unity"). These neo-Pythagorean systems explored how that unity could lead to multiplicity, and postulated multiple metaphysical layers from the topmost being down to material reality. They depended upon numbers, both for terminology and for conceptual patterns. Well into the second century, this strain of neo-Pythagoreanism—a literary ideal, not a community—coexisted with traditional Platonism. Dualism, for example, continued to be championed by philosophers such as Plutarch, who saw it as the normative motif in all classical philosophy.[1] Both monistic and dualistic systems are reported by the skeptic and anti-Platonist Sextus Empiricus (ca. 160–210).[2]

The most vivid witness to the encroachment of monistic upon dualistic territory is provided by Numenius, a second-century Platonist philosopher who flourished in Apamea. Like other Platonists of his day, Numenius was firmly committed to the philosophical importance of numbers. He wrote a treatise called *About Number*, and he preferred a numerical to a geometrical explanation for the composition of the soul.[3] Numbers had religious importance for him as well. He envisioned a person's return to perfective unity as starting with the mathematical disciplines, continuing with contemplation about numbers, and terminating in the final μάθημα, that of being.[4] This religious approach to numbers was interlocked with his view of the Pythagorean tradition. In his view, Pythagoras called God by the name of Monad, and matter he called Dyad. The dyad was in no way generated from the monad; it was completely inoriginate. Numenius castigates "some Pythagoreans"—here setting recent followers against the master himself—for misunderstanding the dogma, claiming that the dyad was derived from a unique monad when the monad withdrew from its nature and sojourned into the condition of a dyad, "an absurd situation, that that which was Monad should cease to be so, and that the Dyad which had no existence should come to subsist, and that thus Matter should come to be out of God, and out of unity immeasurable and limitless duality."[5]

---

[1] Opsomer 2007:381, 389–390, 395.

[2] Sextus Empiricus *Against the Logicians* 10.276–283 (dualistic), 10.281 (monistic); *Outlines of Pyrrhonism* 3.153–154 (monistic).

[3] Frags. 1c, 39.

[4] Fragment 2.

[5] Sed non nullos Pythagoreos uim sententiae non recte assecutos putasse dici etiam illam indeterminatam et immensam duitatem ab unica singularitate institutam recedente a natura sua

Despite his hostility, Numenius was not impervious to the winds of change. His signature doctrine is a three-tiered hierarchy of principles. The first principle, which he regards as an ineffable entity, participates in, but remains above, being. Opposite this is matter, the dyad. The second level of his system involves being, which is being in motion or the demiurgic Intellect. Being comes into contact with matter, and the encounter results in opposite effects. On the one hand matter becomes unified; on the other hand, the contact with matter splits the second principle, giving it a lustful component. Out of the bifurcated second principle comes the third, which governs the material world.[6] Thus Numenius' system has incorporated gnosticizing overtones, by associating the natural world with the lustful activity of the Demiurge.[7] And he has embraced the metaphysical number symbolism and the multiple tiers of neo-Pythagoreanism.

The final step in the transformation from Platonist dualism to monism came in the early third century, through Plotinus (205–269/70), whose writings were so deeply influential that subsequent philosophers are frequently called neo-Platonists, as distinct from the Platonists before him. Because his literary corpus is preserved much better than that of nearly any other ancient philosopher, it is easy to treat Plotinus as a thinker *ex nihilo*. But he was deeply influenced by philosophers of the preceding generation, particularly Numenius and Plotinus' teacher of eleven years, an avant-garde Alexandrian named Ammonius Saccas (fl. early third c.). Echoes of Numenius' doctrines can be seen in Plotinus', and the unwritten doctrines of Ammonius served as the basis for the first ten years of Plotinus' own teaching career.[8]

Number plays such a critical role in Plotinus' philosophy that the topic consumes studies far more extensive than this one.[9] Generally speaking, he preserves the tendency among the neo-Pythagoreans to incorporate number symbolism into metaphysics. He holds to a universe consisting of four metaphysical levels. At the very top is the One or Good, which is beyond all being and beyond all predicates. Below the One is the level of Being or Intellect. And below that is the realm of Soul, followed by matter. Number features on all four of these levels, as pointed out by Svetla Slaveva-Griffin:[10]

---

singularitate et in duitatis habitum migrante—non recte, ut quae erat singularitas esse desineret, quae non erat duitas subsisteret, atque ex deo silua et ex singularitate immensa et indeterminata duitas conuerteretur (fragment 52, trans. Dillon 1996:373).

[6]   Fragment 11; Dillon 2007:366–374.
[7]   Previously noted at Dillon 2007:369.
[8]   Numenius: Dillon 1996:366, 372, 374, 381. Ammonius Saccas: Porphyry *Life of Plotinus* 3.
[9]   See Nikulin 2002 and Slaveva-Griffin 2009.
[10]   2009:87.

Being	Unified number
Intellect	Number moving in itself
Complete Living Being	Encompassing number
Beings	Number unfolded outward

These four metaphysical levels, and the prominence of various levels of "one," are reminiscent of the Pythagorean metaphysical system reported (but not explicitly embraced) by Moderatus around a century and a half earlier.

Plotinus' monism dominates his writings, providing a conduit to his philosophical use of number. For example, he describes the One as "the maker of number. For number is not primary: the One is prior to the dyad, but the dyad is secondary and, originating from the One, has it as definer, but is itself of its own nature indefinite."[11] There is no room for a coeval dyad, as found in Plutarch's and Numenius' writings. Plotinus even declines to discuss at any length the nature and disposition of the indefinite dyad, which according to Aristotle had played such a key role in Plato's doctrines.[12] The monistic emphasis harmonizes with Plotinus' conception and definition of number. He acknowledges, and even uses, the Aristotelian definition of number, but he holds primarily to the generative definition.[13] He is the first philosopher we know of to take the notion of the One's unfolding into number and reverting back to the One—a notion that underlies the definitions embraced by Moderatus and Nicomachus—and extensively integrate it into a major philosophical system. But unlike his Pythagoreanizing predecessors, as well as Valentinians and other gnosticizing groups, Plotinus prefers the term 'hen' to 'monad.' When he uses 'monad' and its cognates, he normally does not differentiate it from 'hen,' and sometimes he conflates the words.[14] So although he adopted many of the philosophical elements of the Pythagorean approach to Platonism, he preferred non-Pythagorean terminology.

Gnosticizing Christians played an important role in shaping Plotinus' philosophy of number. They were quite active in Plotinus' circle, and he and his

---

[11] ὁ τὸν ἀριθμὸν ποιῶν. Ὁ γὰρ ἀριθμὸς οὐ πρῶτος· καὶ γὰρ πρὸ δυάδος τὸ ἕν, δεύτερον δὲ δυὰς καὶ παρὰ τοῦ ἑνὸς γεγενημένη ἐκεῖνο ὁριστὴν ἔχει, αὕτη δὲ ἀόριστον παρ' αὐτῆς· (Plotinus *Ennead* 5.1.5.5–7, trans. Armstrong 1984).

[12] *Ennead* 5.1.5 and 6.7.8 are, to my mind, the only clear examples. Plotinus uses δυάς mostly of duality in general, as a correlate term to μονάς, τριάς, and δεκάς, to explore sets of multiple items.

[13] Plotinus *Ennead* 6.3.12, Slaveva-Griffin 2009:49–50.

[14] *Ennead* 6.2.9.16–18. See also Nikulin 2002:77; Slaveva-Griffin 2009:92–94.

students devoted considerable industry to refuting their positions.[15] One of these refutations is preserved in the last quarter of Plotinus' lengthiest work, a treatise Porphyry titled *Against the Gnostics*.[16] In this treatise Plotinus, referring to an unnamed "they," attacks "their" proposition that there may be more than three principles in the universe, the sort of proposition a Valentinian of his day might have inferred or taught. He writes about this topic only briefly, concentrating instead on the gnosticizing assaults on the goodness of the Demiurge and the created world. But the problems Plotinus discerned in Christian gnosticizing groups, problems that surface at the beginning of *Against the Gnostics*, seem to have remained important to him, because the very next treatise he wrote was his principal treatise on the subject, *On Numbers*.[17] That treatise opens with the same topic that closed *Against the Gnostics*, namely a consideration of evil. It then turns to the question of whether multiplicity and deviation from the One—and therefore we—constitute an evil. The work moves quickly from its original impetus and on to a meditation of what numbers are and where they stand in the metaphysical order. It is one of Plotinus' densest and most difficult treatises.

Plotinus' discussion takes up from Aristotle's criticisms of Plato's doctrine of mathematics, and attempts to broker an agreement, one that would resonate with his third-century audience.[18] He distinguishes, in the realm below the One, between substantial number (οὐσιώδης ἀριθμός, described in terms of monads) and monadic number (μοναδικῶς ἀριθμός, described in terms of henads). The former exists in the realm of being; the latter, in multiplicity. Of the two classes, he is most interested in the constitution of higher, substantial numbers, which are arrayed yet further to describe the different tiers of the intelligible realm. Being corresponds to unified number; beings, to unfolded number; intellect to number moving in itself; and complete living being to encompassing number. These four levels of number correspond to the unfolding of geometrical shape, from point to line to circle to sphere—that is, a foursome similar to Moderatus' first system, and reminiscent of the neo-Pythagorean use of the τετρακτύς.[19]

One of the main points Plotinus argues for echoes his anti-"gnostic" theme: there is beauty and majesty even in the realm of plurality. That is, the realm beyond the One is not all decrepit. This need not be interpreted as a rebuke of gnostics such as the Valentinians; it could be a clarification, or even an affirmation of their insight. Whatever his disposition to gnosticizing Christian number

---

[15] Porphyry *Life of Plotinus* 16.
[16] *Ennead* 2.9. On the kinds of gnosticizing Christians Plotinus was refuting, see Turner 2006:26–27.
[17] *Ennead* 6.6; Porphyry *Life of Plotinus* 5.
[18] For Plotinus' philosophy of number, beyond the outlines sketched in this paragraph, see Slaveva-Griffin 2009:85–99, 122–130.
[19] Griffin 2009:122–126. See also pp. 52–58 above, and p. 185 below.

symbolism, it catalyzed his own constructive explanation for where number sits in his metaphysical edifice. The highest levels of plurality in the early Christian theology of arithmetic were scions and roots of harmony. The Valentinians had idealized the Pleroma as a place of beauty (barring Wisdom's error). Plotinus' philosophy of number explores the rational explanation for this metaphor.

Through this treatise Plotinus strengthened in the Platonist tradition a reliance upon number in all sorts of ways. His literary executor, Porphyry (234–ca. 305), used number symbolism to arrange Plotinus' oeuvre.[20] And Amelius (b. 216–226, d. ca. 290–300), a philosopher who, according to clues in the *Life of Plotinus*, may have had greater respect from the master than Porphyry had, repeatedly developed new ideas about the soul and its relationship to number.[21] He reportedly wrote a forty-book response to the Sethian treatise *Zostrianos*.[22] One of Amelius' innovations was to use the numerical patterns laid out in Plato's *Timaeus* as a template for classifying the various kinds of souls in the world.[23]

This trend took a new turn in Theodore of Asine (ca. 275–ca. 360), a philosopher who perhaps met Amelius in the 290s and studied with Porphyry in the opening years of the fourth century.[24] He found his inspiration in the doctrines of Numenius and Amelius, and garnered a following of at least modest size. Toward the end of Theodore's life, his followers were still active, arguing on behalf of their master. Of his writings we know only two titles, *On Names* and *That the Soul is All the Forms* (sc. *of Life*), extracts of which, some lengthy, are preserved by Proclus.[25]

Theodore's metaphysical system, as obscure today as it ever was, involves among other things a complex explanation of the origin and structure of the soul, the same theme that Amelius developed. But he bases his metaphysical edifice upon a foundation of letters, numbers, and psephic exercises previously unknown in the Platonic tradition. He explains the soul on the basis of letters, their numerical value, and the symbolism behind the numbers. The bulk of this numerically oriented system is preserved in testimony 6, worth quoting in full.[26] Proclus, our source for the testimony, says:

---

[20]   See Slaveva-Griffin 2009:132–140.

[21]   On respect: Amelius kept Plotinus' notebooks, he was given authority by Plotinus to refute Porphyry, he was too busy to correct copies of Plotinus' lectures that Porphyry requested, and he was listed by Longinus as one of the two chief Platonists teaching in Rome (the other being Plotinus): Porphyry *Life of Plotinus* 1–5, 7, 16–21. On Amelius in general, see Brisson 1987.

[22]   See Turner 2006:26–27, esp. n. 19.

[23]   Stobaeus *Anthology* 1.49.37.21–24, 1.49.38.1–4 and Brisson 1987:838–839.

[24]   Brisson 1987:818–819.

[25]   On Theodore's life see Brisson 2002; PRE 5 (n.s.): 1833–1838 (Theodore, no. 35); Gersh 1978: 289–304.

[26]   Theodore's fragments are collected in Deuse 1973. Testimony 6 comes from Proclus *Commentary on the "Timaeus,"* in Festugière 1966–1968:2.274–278. For the Greek text, see the Appendix, p. 199.

Theodore, the philosopher of Asine, inspired by the discourses of Numenius, in a rather novel manner composed treatises concerning the generation of the soul, making his attempts from letters, written characters, and numbers. So that, then, we might have concisely his written opinions, come, let us make an overview, point by point, of everything he says.

[1][27] So then, the first is rightly hymned as being, for him, ineffable, inexpressible, fount of all things, and cause of goodness. [2] And after this [first], so exalted above all things, there is a triad that defines, for him, the intelligible plane. [The triad] he calls the One [ΤÒ ῎ΕΝ] since it comes [2a] from a breathing that is somehow ineffable (which the rough breathing of ἕν mimics), [2b] from the loop of the Ε itself, on its own without the consonant, and [2c] from even the Ν itself. [3] Another triad after that one defines the intellectual depth, [4] and another one, the demiurgical. The former is existence prior to being, thinking before mind, living before life. After these is the demiurgical triad, containing first, being; second, mind; and third, the fount of souls.

[5] From that triad is another triad: absolute soul, universal [soul], and [soul] of the all. We have earlier discussed the distinction of these things, each of which proceeds from the entire demiurgical triad—that is, one from being, another from mind, and the other from the fontal soul.[28] Indeed, it was proposed that Plato speak about this soul of the all, especially about the plain soul that comes from the fount of souls, about the universal [soul] and the [soul] of the all, and about the fount itself. For all things exist in all things, even if at one time in one way, and at another time, another:

> in the soul before the triad [all things exist in all things] in
>   unity;
> in the plain [soul], [all things exist in all things] in wholeness
>   before the parts;
> in the universal soul, [all things exist in all things] in wholeness
>   from the parts;
> and in the third [soul], [all things exist in all things] in [wholeness] in the parts.

[All this is said] on the basis that Plato classified all these things and needed to refer to every [soul] every ratio, ratios that allow no difference among them.

---

[27] My numeration and lettering appears in square brackets to mark passages for discussion below.
[28] This refers to a discussion preserved in test. 22.

And [Theodore] thinks it necessary to say first why [the soul] exists from three means. And indeed he says that the soul as a whole is a geometrical ratio, existing from both the first god according to being and the second [god] according to mind. For these very things are two essences, one undivided and the other divided. Both the arithmetical ratio (which bears the image of the first essence) and the harmonic [ratio] ([which bears an image of the] second [essence]) result in [the geometrical ratio]. The former is monadic, since it is without extension; the latter is discrete, but harmonically so.

[7] Then, the entire number might be a certain geometrical number, since [the soul] is shown to be a tetrad, being from the tetrad of the elements/letters. But lest you suspect that this number is lifeless, taking for the third heptad the first, you will find life in the extreme letters.

Rather, setting out according to its order the base of the first letter, you will see the soul is an intellectual life. E.g. Z, O, Ψ. The middle [term, O,] is the circle, being the intellectual one, since mind is the cause of the soul. The smallest [term, Z,] shows [that the soul is] geometrical, a kind of mind, through the attachment of the parallels and the diameter. [The mind] remains above and encompasses opposites and is shown to be a form of life, both oblique and not oblique. The largest [term, Ψ,] is the element/letter of a sphere. So then the lines, bent into each other, will make the sphere. On top of this, the next letter's bases, Δ, M, Y, are simultaneously three <u>and</u> tetradic. And because of this, as they beget the dodecad, they result in the twelve spheres of the all. The largest of the bases[, Y,] shows that its essence yearns for two certain things and stretch up toward two affairs. Therefore some call this letter 'philosopher.' The [essence] of both flows to the lower [region]. So this is why we find the Y referred to by some of the noteworthy [authors of the past]. It is between two spheres, the Ψ and the X [i.e. in the word ΨΥΧΗ], the former being warmer (because of the breath) and more life-giving, and the latter having each [quality] to a lesser extent. Thus, there is again a mean between two minds, one earlier and the other later, and the middle character makes clear its property and relation to the other. Rather, even though the letter Ψ is a sphere, Plato assigned the X to the soul, so that it might show the equal balance of motion itself, since all the lines in the X are equal, and thus to make the automation of the soul evident. But if the Demiurge brings in the soul through existence itself, it is clear that he himself has ordered it in proportion to the X. After all, that is the foremost mind. And so, because of these things, he says that the soul, as it advances and brings itself out, is a certain middle essence of two minds. And this is the manner in which these things are to be understood. But through the last letter, the H, the advance of [the soul] up to the cube is to be observed.

[8] And if it is a dyad because of the otherness of life, and it is a triad because of the tripartition of its essence, then it has, on the one hand, the ratio 3:2. But as it enters into itself and, through its entrance, applies the dyad to the triad, it begets the hexad. As it connects to the undivided and to the trisected the harmony that [comes] through these things in doubleness, it comes into existence. And since, on the one hand, the triad, as it turns into itself, results in the ennead, and the dyad, on the other hand, moving into itself dyadically always results in the octet, so from both it results in the ratio 9:8.

The linear birth [of the soul] makes clear its indivisibility, and its thorough homogeneity (after all, every part of a line is a line), and that all the ratios are everywhere. But the split into two demonstrates that its form is dyadic. And its indivisible wholeness is an image of the first mind, whereas the unsplittable [wholeness] of the two (which he calls the circle of the same) [is an image] of the second [mind]. And the [wholeness] split into six [is an image] of the third [mind], the last to be calculated. And the octet becomes manifest from the dyad of the soul, whereas the heptad depicts in monads the first form of the soul; in decads, its intelligible [form] (because of the circle); and in hundreds, the soulish mark, the third one remaining. And [the soul's] straight, connate nature exists for the fixed [sphere], which begets; whereas the exit and indefinability [exist for] the wandering [sphere]; and the return after the advance [exists for] the life that wanders without wandering. And since on the other hand the shape of the soul is like X, and its form is dyadic (since the split is in two), and the dyad [applied] to the hexad (being primarily the base of X) creates the dodecad, you might take from that the first twelve ancient souls.

Although the testimony continues, it is worth commenting on a number of ideas that chaotically emerge. Theodore's system reaches a new level of complexity, similar to the intricacies of Marcus' theology. It usually takes several slow readings to understand much of Theodore's system. Proclus' summary is terse, and even in faithful translations many ideas are incomprehensible.[29]

---

[29] For other translations, of varying degrees of accuracy, see Proclus *Commentary on the "Timaeus,"* trans. Festugière 1968:3.318–321; Ferwerda 1982:51–52; Funk et al. 2000:214–216. Taylor, the great nineteenth-century English Platonist, skipped it, explaining, "Proclus gives an epitome of

At the top of Theodore's system (marked in the translation by [1]) is a transcendent entity, described in negative terms that describe its role as cause and source. Significantly, he does not call it "One." Rather, it is absolute, beyond names, beyond even spirit or breath.[30]

Below this transcendent entity is the intelligible plane [2], consisting of a triad that unfolds and manifests the primal entity in the letters of ἕν. The rough breathing [2a], which Theodore claims is silent, mimics the ineffable breathing of the uppermost entity.[31] This ineffable breathing is symbolized by the rough breathing, the δασεῖα, which had not been pronounced for quite some time.[32] The unspoken δασεῖα is the perfect symbol of the ineffable, since it embodies the paradox of unspoken speech. According to another of Theodore's testimonies, its combination with the Є, shaped by a loop [2b] (remember, Theodore is commenting on uncial Greek letters, not minuscule) renders the vowel at once utterable and unutterable. The N [2c] rounds out the image of the ineffable made effable.[33] The triad 'ЄN consists of two groups: a dyad (presumably ЄN) and, above it, a monad (presumably ').[34]

Beneath this level is the intellectual depth [3], consisting of a triad: existence, thinking, and living. Each of these three parts of the intellectual depth stands over the elements in the triad just below, the demiurgical depth [4], a triad that consists of being, mind, and the fount of souls. This fourth level seems to have been an easy target for Theodore's critics, who saw this doctrine, taken or derived from Amelius, as unnecessarily introducing three demiurges, in lieu of Plato's one.[35] The threefold demiurgical level, however, sustains Theodore's commitment to a metaphysics of triadic descent. It also critically explains the next level down, the realm of the soul [5], which consists of three kinds, all derived from the demiurgical triad in general, and the fount of souls in particular.[36] Each of the three aspects of the soul corresponds to a certain relationship between the whole and the parts, and to the three major kinds of ratios. What corresponds to what is treated in yet another testimony, where Theodore goes to

---

this theory, but as it would be very difficult to render it intelligible to the English reader, and as in the opinion of Iamblichus, the whole of it is artificial, and contains nothing sane, I have omitted to translate it" (Taylor 1820:141n1).

[30] Pace Funk et al. 2000:214, 216, 227. It is inaccurate to call it the One because in test. 9 (Deuse), Theodore explicitly denies that this highest entity has a name.

[31] Test. 9: *sicut et spiritus tacitus* and *spiritum ... indicibilem.*

[32] Gignac 1976:134–135.

[33] Test. 9 is unclear. See Morrow and Dillon 1987:590 and n. 113. Turner's suggestion (Funk et al. 2000:216) that 'ЄN represents point, line, and plane reads too much into the passage.

[34] Test. 9 neither stipulates which letters are the monad and which the dyad, nor elaborates on how the monad generates the dyad.

[35] See test. 12, translations by Taylor 1820:260; Funk et al. 2000:218–219; Festugière 1968:4.165–166.

[36] For amplification on this level, see test. 22, which is translated in Funk et al. 2000:221–223, Festugière 1967:3.262–265, and Taylor 1820:92–94.

some lengths to correlate the numbers one, two, three, four, and six with the four elements, and by extension with the soul.[37] For example, the series one, two, four, a geometric series ($b/a = c/b$), corresponds to Earth because of its name, and so the number seven, the sum of the series, also represents the Earth.

Theodore's complex number symbolism, linking souls to the sublunary region, shows that he saw numbers as a key constituent in the formation, design, and explication of the universe. The most basic parts of this number symbolism are evident in the structures of his metaphysics, presented in testimony 6 [1–5], which may be diagrammed:

	τὸ πρῶτον ἄρρητον 'the ineffable, first'		
νοητὸν πλάτος 'intelligible breadth'	ʽ O	Ε N	N E
νοερὸν βάθος 'intellectual depth'	εἶναι 'existence'	νοεῖν 'thought'	ζῆν 'life'
δημιουργικὸν βάθος 'demiurgical depth'	ὄν 'being'	νοῦς 'mind'	ζωή, πηγὴ τῶν ψυχῶν, πηγαῖα ψυχή 'life', 'fount of souls', 'fontal soul'
	αὐτοψυχή 'absolute soul'	ἡ καθόλου ψυχή 'universal soul'	ἡ τοῦ παντὸς ψυχή 'soul of the All'

Here is vividly portrayed Theodore's mathematically molded view of the divine realm.[38] The highest realm of the universe is divided into thirteen places: four planes, each composed of a triad, all governed by a transcendent thirteenth entity. Theodore's system is unusual, but not unique. Other Platonist texts, such as the Nag Hammadi text *Marsanes*, contemporary perhaps with Amelius, agree in the number of levels and in the presence of an ineffable, transcendant entity located in the thirteenth place.[39]

---

[37] Test. 22. See n. 36 above.

[38] See comparative schemes at Deuse 1973:22–24, Funk et al. 2000:230, Finamore 2000:256–257, and Edwards 2006:122–124.

[39] But prominent differences should also be noted. Marsanes' lowest realm is the corporeal world, not the soulish. And even though Marsanes teaches that the twelve stages culminating in the thirteenth consist of four levels of triads, those triads do not follow the consistent pattern found

Later in testimony 6, Theodore points to the soul as consisting of four elements or letters [7]—the play on the double meaning of στοιχεῖον is evident—as the reason for calling the entire number of the soul a geometrical number. This introduces a letter-by-letter, numerical explanation of the word ΨΥΧΗ, which is geometrical since it consists of four letters. Theodore's method here focuses on the number of letters in a word, not their psephic value.[40] He takes a letter, for example Ψ, finds its base (πυθμήν), then lists other letters that share the same base, in this case, Z and O.[41] That is, Ψ, O, and Z have values of 700, 70, and 7. He then uses each of the three letters, particularly its shape, to explain the letter as a whole. The combination of parallel and oblique lines in the letter-form Z explains how the soul preserves both kinds of directions, as specified in Plato's *Timaeus*. The circle of the O represents the mind's generation of the soul. The Ψ's crossarms, arching in toward each other, represent a sphere. For the Y in ΨΥΧΗ, its base of four, when multiplied by the three letters that share it as a base, yields the "twelve spheres of the all," presumably the zodiac. Its shape too illustrates the philosophical implications of the letter, as its stem and arms portray the flow from upper regions to lower, and the return. The two arms of the Y stretch toward its companion letters, Ψ and X, the former being more life-giving (as Theodore stipulates elsewhere, Ψ recalls Z, the initial of ΖΩΉ 'life'), the latter more soullike.[42]

The final part of the testimony [8] deals with the *Timaeus*' double and triple progressions and identifies the activities of soul with the various numbers that make up the most basic musical intervals (e.g. 3:2 and 9:8, the perfect fifth and the whole tone).[43] The passage is too complex to explore here. Nevertheless, note how deeply Theodore's explanation of the constitution of the world soul involves itself in arithmetic. Theodore is squarely in the center of Pythagorean Platonism, which identified number, combined with motion and the mixture

---

in Theodore: existence-thought-life. The regularity of Theodore's triads lends itself to depiction in a rectangle, where both horizontal and vertical relationships are important. Theodore, for instance, places living over life (the two terms are cognates), but their corresponding entities in *Marsanes*, self-generated power/triple perfect and super-(corporeal?) do not have this kind of vertical relationship. *Marsanes*' thirteen seals are presented seriatim.

[40] See below for Iamblichus' criticism, which confirms that Theodore's method is about letter counting, not psephy. Sambursky 1978 suggests that the term 'geometrical' in test. 6 preserves the earliest use of the term 'gematria.' But what very basic psephy there is has nothing to do with what Theodore calls the "geometrical number," a term that refers to the concept of geometrical ratio, discussed just before.

[41] Theodore's use of πυθμήν to refer not merely to the base but to its multiples of ten and one hundred goes against normal use of the word. Ordinarily O and Ψ would have a single πυθμήν, Z, and would not themselves be πυθμένες. See LSJ, s.v., and the psephic numerology attested in Hippolytus *Refutation of All Heresies* 4.13–14.

[42] On the Y as a Pythagorean symbol of moral choice see Harms 1975 and Dornseiff 1925:24.

[43] Plato *Timaeus* 35b: 1, 2, 4, 8 and 1, 3, 9, 27.

of same and other, as the source of the soul's origin.[44] Theodore indulges in this tradition, and expands the application of numbers, making arithmetic a constituent part of every level of his philosophical system. He introduces letter symbolism to the interpretation of the *Timaeus*. In this he resembles Monoïmus, who took the shape of the ἰῶτα as a reflection of the divine. He has even more in common with Marcus, who introduced letter symbolism into Valentinian protology. Marcus, for example, identifies the number of letters in the name 'Jesus Christ' as a symbol of the structure of the aeonic realm; so Theodore treats the four letters of ψυχή as cosmologically significant. Theodore finds the shape of the letters symbolizing important metaphysical truths, just as Marcus locates in the divisions of the alphabet signs of Pleromatic emanations. Both indulge in rudimentary psephy, revealing a common assumption that words, names, and numbers have an interconnected significance well beyond what is immediately apparent. Analyzing the psephic value of a word's letters allows one to discover hidden knowledge, and to learn more about the universe or about the person or thing that bears the name.

Just as Marcus' speculation became a prime target for Irenaeus, so Theodore's earned him the reproach of Iamblichus (ca. 245–ca. 326), his older contemporary and rival Platonic philosopher. Iamblichus, quite possibly a relative or descendant of Monoïmos,[45] had also studied with Porphyry. He later established a philosophical circle in Syria, one that Theodore eventually joined as his student.[46] Iamblichus' criticism is preserved in the last part of Theodore's testimony 6. Proclus says:[47]

> Thus Theodore philosophizes these kinds of things about these matters, making his interpretations out of letters and utterances (to compare a few [ideas] among many). But the divine Iamblichus lambasted this sort of viewpoint in his responses, *Against the Circle of Amelius* (for so he titles the chapter) and indeed also [in *Against the Circle of*] *Numenius*. [Iamblichus] either—for I cannot say [which]—identified [Theodore] with these [two men] or had somewhere found them writing similar things about these matters.
>
> So the divine Iamblichus says first that you shouldn't make the soul the entire number or the geometrical number because of the quantity

[44] This interpretive tradition of the *Timaeus* began with Xenocrates. See Plutarch *Genesis of the Soul in the "Timaeus"* 2, 22.
[45] See Dillon 1987:865 for analysis of a medieval report (Photius *Biblioteca* 181 [126a]) that Iamblichus had ancestry from Sampsigeramos (fl. 60 BCE) and Monimos.
[46] Eunapius *Lives of the Sophists* 458. On Iamblichus' life, see Dillon 1987:863–875.
[47] In addition to those listed in n. 29 above, other translations are in Dillon 1973:165–167 and Taylor 1820:141.

of its letters. For ΣΩΜΑ ['body'] too is made of the same [number] of letters, as is even ΜΗ ΟΝ ['nonbeing']. So [by Theodore's reasoning] ΜΗ ΟΝ would be the entire number. You could find many other things that consist of the same [number] of letters yet are shameful and completely opposite to each other, all of which would certainly not be right to conflate and confound with each other.

Second, it is dangerous to try [to build a system] based on written characters. After all, these things are relative: the carving of an archaic [character] used to be one way, but is now another. For instance, the Z upon which that man builds his argument had neither parallels that were completely opposed, nor the middle diagonal bar. Rather, [the crossbar was] perpendicular, as is apparent from ancient stelae.

Third, to reduce [numbers] to their bases and to preoccupy oneself with them, [going] from one number to another or vice versa, alters our understanding. For the heptad in monads, the [heptad] in decads, and the [heptad] in hundreds are not the same thing. So if this [heptad] was in the term ΨΥΧΗ, why must he sneak in an account about bases? After all, using this same technique we might transform every thing into every number, by dividing or adding or multiplying.

So much for the general [problems]. [Iamblichus] also refutes [Theodore's] individual results, [showing them to be] fraudulent and insane. And everyone who would like to know the weakness of every point may easily acquire the treatise and read through the appropriate refutations of everything [taken] from [Theodore's] writings.

We see in this passage a quote from a quote from a quote, and it may be helpful to sort out what is before us. Proclus, our source for testimony 6, had before him a work by Iamblichus, a book of polemic divided into chapters discussing and arguing against a series of philosophical schools.[48] One chapter of that book was directed against the circle of Amelius, and another against that of Numenius. Proclus himself is in the dark about what he is reading. He is uncertain whether Iamblichus meant to identify Theodore with these two circles, or merely to refute an approach shared by all.[49] This suggests to me

---

[48] Here I agree with Festugière against Dillon (1973:337), who understands the text to refer to one title (*Against the Circle of Amelius and Numenius*), not two (as in my translation). The two chapters and their titles, described as refutations, show that the source is probably not a commentary on the *Timaeus*, as Dillon surmises, but a critical summary of various philosophical schools of thought.

[49] My interpretation depends upon taking the τοῦτον of Festugière 1966–1968:2.277.30 with Theodore, and the ἐκείνους of the same line with the circles of Amelius and Numenius. Dillon's reading (1973:165, 337–338), which assigns τοῦτον to Numenius and ἐκείνους to the circle of Amelius, is possible. But the overall context of Proclus' report puts Theodore front and center,

that Iamblichus did not name Theodore specifically in his polemical work, but refuted Theodore's doctrine under the name of these two schools, leaving the reader to make the connection.[50] In any case Proclus, who knew Theodore's writings well, recognized that Iamblichus' critique applied to Theodore.

Iamblichus makes three criticisms of Theodore's method in general, and then refutes some specific doctrines. Proclus reports only the general criticisms. The first is Iamblichus' contention that one cannot infer from the premise " 'soul' consists of four letters" that the soul is the entire number, or that it is the geometrical number. This first criticism deals with Theodore's fascination with the number of letters in ΨΥΧΗ. His critique, Iamblichus says, has nothing to do with the practice of gematria, but with letter counting.[51] He demonstrates the faulty logic by deploying a *reductio ad absurdum*. Both 'body' and 'non-being' have four letters, so Theodore's method should allow one to declare a similar, exalted position for either corporeality or nonexistence. Iamblichus notes that a number of four-letter words, too rude to mention (he would have appreciated our English idiom), could be inserted in Theodore's system. Would that make them just as exalted as 'soul'?

The second criticism is that letterforms are a faulty starting point because of their relative nature. Iamblichus focuses on how letterforms change, demonstrating it with the Z. He points out that the archaic form of the letter was I, a shape that undermines Theodore's interpretation of the soul's being at once parallel and oblique.

Third, Iamblichus criticizes as valueless the practice of taking alphabetic numerals, finding their base value, and extrapolating from that base to other numerals with the same base. He questions why bases should even be a point of consideration. The risk in the practice is that someone could take a word like 'soul' and derive any preconceived result, through arbitrary mathematical operations. Iamblichus may be somewhat unfair here to Theodore, who follows a principle common in psephic games, that one may move from one number to another by dividing or multiplying by ten. But Iamblichus correctly criticizes the practice for its attempt to achieve preconceived results. The technique could be used to prove whatever one desired.

Now, Iamblichus was no enemy of number symbolism. His writings are full of Pythagorean arithmology. And he was the first Platonist to use number symbolism to organize his own literary output. His magnum opus, *On Pythagoreanism*, occupies ten books, to pay homage to the perfection of the number

---

much closer (and therefore τοῦτον) than Amelius and Numenius (here only obliquely referred to through treatise titles, and so ἐκείνους).

[50] See Dillon 1973:338.

[51] Contra Dillon 1973:338–339 and Sambursky (see n. 40 above).

ten.[52] And each of the extant books in this series deals in some way with number symbolism, if not arithmetic. In the third book, *Common Mathematical Knowledge*, for instance, Iamblichus argues for a theory of mathematics. He defends Pythagorean mathematics as useful for sharpening one's understanding and for purifying the soul in preparation for union with the divine.[53] His metaphysics was built on mathematics just as Plotinus' was, and he had a system of triads that resembled Theodore's (a triadology characteristic of subsequent Neoplatonism).[54] Iamblichus' numerical interests were so well known that he was credited with a handbook of number symbolism popular throughout the Byzantine period, the otherwise anonymous *Theology of Arithmetic*.[55]

Like Theodore, Iamblichus engages in extensive symbolic analysis of the numbers in the *Timaeus*. Both men explain the generation of the soul using arithmetical explanations for the generation of the dyad and triad from the monad.[56] So Iamblichus criticizes Theodore, not for number symbolism per se, nor even for extravagance (that being a matter of taste). His main complaint is that Theodore's method is capricious, bound to changeable human convention, and liable to lead one to all sorts of absurd conclusions. Theodore's method destroys the very essence of a symbol. Iamblichus prizes numbers as symbols because of their inherent connection with higher realms of reality. Theodore focuses, however, on facets of numbers and letters that are rooted in the human condition, the product of changing social conventions. In these contexts the symbol loses its power because it depends upon temporal realities, not eternal ones. To locate numerical symbols in the realm of the divine was completely justified and expected in the Pythagorean and Platonic traditions. But to root them in human convention, as Theodore did, was not.

The debate between Iamblichus and Theodore compares favorably with that between Irenaeus and Marcus. Both Iamblichus and Irenaeus deploy biting sarcasm in their *reductiones ad absurdum*, to discredit an approach that threatens to disregard the rules for attaining truth. Both writers criticize techniques that

---

[52] For the structure of this work, see O'Meara 1989. On the perfection of ten, see p. 54n78 above. For examples of Iamblichus' symbolism beyond those given here, see Shaw 1999.

[53] Iamblichus *Common Mathematical Knowledge* 18.7–23.

[54] Dillon 1987:883, 888–889.

[55] No internal evidence associates the *Theology of Arithmetic* with Iamblichus, but the substance of the treatise, with all its theological and scientific number symbolism, fits well with his general outlook. Tarán 1981:291–298 argues that the text, dated traditionally to the fourth century, is probably the product of a later Byzantine compiler who drew from arithmological treatises of Nicomachus of Gerasa, Anatolius of Laodicea, Iamblichus, and others. O'Meara 1989 overturns some but not all of Tarán's assumptions. The date and textual history of the *Theology of Arithmetic* needs to be revisited.

[56] Proclus' *Commentary on the "Timaeus"* depends considerably on Iamblichus. See Dillon 1973:160–163, 322–325.

are custom-bound and apt to lead a person to preconceived ideas, or to absurd, immoral positions. Both Irenaeus and Iamblichus participate in communities grounded in tradition, and they both work with systems and texts that were considered to have roots, so to speak, in higher, divine soil. For Irenaeus, this is the apostolic rule of faith, deposited in the churches and codified in the Bible; for Iamblichus it is the Platonic rule of faith, as practiced in religious theurgy and enshrined in the writings of Plato. The logic had similar effects in both traditions: just as, after the third century, number symbolism was excluded from Trinitarian theology (see below), so in the Platonist tradition the use of gematria and letter symbolism to explicate the highest metaphysical levels ended with Theodore's followers. Neo-Platonists followed their orthodox Christian counterparts.

The comparison should not be pressed too far. Christians and Platonists had incommensurate notions of community and tradition. Christians identified themselves as the new Israel, the holy people of God, and they were committed to the Bible, the inspired charter of their faith. Platonists, who had no church to speak of, were intellectual journeymen, maintaining a tradition of discussing the issues discussed by Plato, Aristotle, and others. Nevertheless, both Irenaeus and Iamblichus characterize their opponents as having strayed beyond the acceptable limits of a tradition both sacred and true, and apply nearly identical logic in their critiques. Their targets, Marcus and Theodore respectively, also share a common approach and set of assumptions about the importance of numbers and letters, regarding them as symbols and portals to the divine.

This is not to suggest that Iamblichus was reading Irenaeus and Theodore was reading Marcus. Rather, Christianity filtered the conceptual air that the Platonists breathed. The theological issue about the appropriate use of numbers first arose in Christianity. The trends and tensions discernible in texts written by the Valentinians, Irenaeus, Clement, Monoïmos, and the *Paraphrase of the "Apophasis Megale"* migrated to the Platonist tradition. First, Plotinus responded to the number symbolism of the "Gnostics" by offering a thorough, penetrating defense of the metaphysical priority of numbers. His successors combined that emphasis with long-standing mathematical interests of ancient philosophers to formulate a triadology that dominated Platonist philosophy all the way through Damaskios and the closing of the academy at Athens (529 CE).[57] This later tradition built upon Iamblichus' theology of arithmetic, but excluded Theodore's psephic techniques.

---

[57]   Wallis 1992, O'Meara 1989.

The Later Christian Tradition

And what of that Christian experience? Where did it end up during the late third and early fourth centuries? To answer this, it is important to recall some of the main points already covered.

The Pythagoreanism that emerged in the first century BCE inspired and even provided the intellectual foundation for the theological systems of the Valentinians, the Barbelo-Gnostics, deutero-Simon, and Monoïmus. All these groups make Pythagorean number symbolism a central part of their Christian theology. They begin with an arithmetical array of aeons, multiple beings initially projected by a monad or monad–dyad pair that is the source of all things. From this top tier down to the physical world, various descending levels of reality are projected. The arithmetical edifice and terminology of the Middle Platonists helped the Valentinians describe the emanations of the aeons, their cascade into multiplicity, and the formation of the natural world and the human being. They brought to these metaphysical systems an emphasis on salvation, telling a story of how we can be or have been rescued from the present world and restored to that metaphysical edifice. Many aeons are given numerical names drawn from Pythagorean lore (e.g. τετρακτύς, 'monad,' and 'dyad') and placed in arithmetical relationships to one another, making explicit the mathematics already implied by the Pleromatic structure. Some systems are philosophically pure, committed thoroughly to monism or to dualism. That is, some focus on the utter solitude of the Monad; others make the eternal relationship between Monad and Dyad central; other systems seem not to have cared. All the models, however, fall somewhere along a monadic–dyadic spectrum. Whether pure or mixed, the different gnosticizing visions all reflected the intellectual concerns of Middle Platonism, in which monism and dualism were live, competing options. This arithmetical theology first developed in the 160s and lasted through the mid-third century. Marcus, who in the 180s fused his theology of arithmetic with grammar and letter symbolism, marks the apogee of that era, an era that coincides, as Christoph Markschies has observed, with a period of highly philosophical and well-developed gnosis.[58]

Middle Platonists also inspired early Christians to incorporate number symbolism into their interpretive techniques. Christians had always been attuned to symbolism and allegory. The exegesis of the Hebrew scriptures developed by various New Testament authors is similar to the interpretive approaches of Philo, Plutarch, and other Middle Platonists. But with the Valentinians came a surge of interest in unlocking the secrets behind numbers found in the Bible. They even used the technique to interpret the natural world.

---

[58] Markschies 2003:85–100.

Everyone—Valentinians, Marcus, Monoïmus, Pythagoïmus, deutero-Simon, Irenaeus, and Clement—read the numbers found in Scripture in light of Greco-Roman number symbolism—the kind espoused by Plutarch and the *Theology of Arithmetic*—to press home their exegetical points. Those numbers could appear in the Bible explicitly, as with the 99 in the parable of the lost sheep, or it could be merely implied, as in the story of the Transfiguration, which itemizes but never numbers those present. Or it could be even more cleverly hidden, in the psephic values of letters or alphabetic numerals, such as the 801 of the word 'dove,' or the numeral 10 in the story of Gideon or the iota not missing from the Law. Christian exegetes latched onto the numbers they read and interpreted them in the light of some other part of their tradition. This impulse developed among orthodox and heretics alike, but in different measures, and to different ends. Clement typifies later, Byzantine number symbolism, for he more than any orthodox predecessor creatively interpreted the number symbolism of the Bible, yet he channeled it to serve orthodoxy, to ensure that the numbers discovered in the Scriptures maintained the traditions of the entire Church, not the private fantasy of an elite group. In this he was more consistent than Irenaeus, who made use of the very exegetical techniques he faulted.

The two interrelated approaches to number symbolism—the philosophical and the exegetical—met different fates. The orthodox theology of arithmetic that emerged in the mid-third century rejected the philosophical, metaphysical use of number, at least one that did not start with the rule of truth. But it permitted an exegete to be as fanciful or reserved as he liked when it came to interpreting the numbers of the Bible.

The first of these two, the demise of number symbolism in descriptions of the godhead, is not difficult to explain. All the systems discussed in chapters 3 through 5 above come off as a form of polytheism. The Valentinians and others like them were telling a story, fundamentally, of how one or two original divine principles begat a pantheon of individual divine entities. These could not be interpreted as mere metaphors, the arbitrary symbols of modern semiotics. They were to be treated as real, and as meaningful as the material world to which they eventually led. This paradigm stood at odds with the monotheism of Irenaeus, Clement, and the church traditions they defended. Irenaeus is unambiguous in his commitment to God as three persons, Father, Son, and Holy Spirit, all separate from Creation. For Clement, the one God remains free of any mathematical models, metaphors, or constraints, including that of the One. For both, neither God nor the Christian tradition is subject to numbers, but rather the converse.

This is not to say that the orthodox were unified in how arithmetic should be used. Irenaeus writes often about one Father and one Son. The oneness he emphasizes steers clear of philosophical metaphors. Clement is more willing to indulge in them. For him, God transcends any predicate, including that of the One. Clement considers numerical analogies fruitful, and he is willing to think about God as a One that stands above whatever else there is below it (including the conceptual hierarchy Monad → One) until the metaphor runs out of steam. He is comfortable with such dispensable analogies, whereas Irenaeus avoids them. The two approaches mark the range of approaches used by Christian theologians throughout late antiquity. Mathematics rarely features in the later Fathers' trinitarian theology, and where it does it is left as an undeveloped simile—one of many—for a God who defied metaphor. Orthodox Christians deployed Platonist terminology or philosophical metaphor subversively, both to draw pagans out of philosophical attachments and into the Church, and to deflect criticism from fellow churchmen for Hellenizing the faith.

The development of number symbolism in later Christian exegesis is also not difficult to explain. Just as Irenaeus and Clement handled numerical symbols differently, in later Christianity there was no single acceptable way to use number symbolism. Those inclined to mystical, speculative, or allegorical theology tended to embrace Clement's pattern; those more skeptical tended toward Irenaeus', or omitted it altogether. The prolific Biblical exegete and theologian Origen (185–ca. 255) frequently uses number symbolism in his interpretation of the Bible, in a fashion somewhat between that of Irenaeus and Clement of Alexandria. A third-century bishop of Laodicea, Anatolius, who may have been the teacher of Iamblichus, collected Pythagorean numerical lore into a handbook, parts of which are preserved in the *Theology of Arithmetic*.[59] His interest and skill in mathematics were fundamental to his rationale in explaining the date of Easter, a logic that proved influential in the fourth century and beyond.[60] The Bible commentaries of Didymus the Blind, Evagrius of Pontus, Jerome, Augustine, and many more are salted throughout with creative number symbolism. Some of their contemporaries, such as Athanasius and John Chrysostom, were not so inclined. But this difference did not lead to controversy or dispute, since all agreed that number symbolism was to be drawn from the tradition, not imposed upon it. This left latitude for those so inclined to creatively develop number symbolism, and throughout the medieval period local traditions flourished in the Latin West, in the Byzantine East, and in the Syriac, Coptic, Ethiopic, Arabic, Georgian, and Armenian traditions. Every one of these regional traditions fostered new kinds of arithmology, collectively indebted to the Greco-Roman tradition and the formative second- and third-century arguments.

---

[59] On Anatolius, see Dillon 1987:866–867.
[60] Knorr 1993:184; Anatolius *On the Computation of the Pasch.*

Figure 7. The Transfiguration of Christ. Göreme, Karanlık Kilise, 11th c.
(Photo courtesy Mustafa K. Turgut.)

A vivid example of the influence of the early Christian theology of arithmetic appears in the Byzantine icon of the Transfiguration. From the earliest, sixth-century examples into the modern period, Orthodox Christians have depicted Christ atop mount Thabor, flanked by Moses and Elijah, standing above the three apostles. Together the six figures form a symmetrical circle or triangle, to emphasize their number and harmony. Behind Christ is a mandorla of light. In the earliest example, a sixth-century fresco from Mount Sinai, the mandorla emits eight beams of light. That eight-rayed motif is found frequently thereafter. In the Palaiologan period the mandorla was arranged as two superimposed rhombuses or squares, their eight corners prominently arrayed (see Figure 7). The eight divine rays are no accident, nor is the human hexad, marking the mountain, as a triangle. Clement's orthodox interpretation of the number symbolism behind the Transfiguration has become indelibly imprinted in the Orthodox tradition, furnishing for this icon number symbols that proclaim Christ as both man and God, the 'episemon ogdoad.'[61]

---

[61] Not all icons show eight rays, but it was certainly the dominant tradition. See Miziołek 1990 and Drpić 2008:241–242. Clement's Transfiguration number symbolism was known in later Byzantium: see Gregory Palamas *Sermon* 34.4–6 (PG 151:425–428).

Excursus A

One versus One

The Differentiation between Hen and Monad in Hellenistic and Late Antique Philosophy

THEON OF SMYRNA'S *MATHEMATICS USEFUL FOR READING PLATO*, written in the second century CE, collects arithmetical, geometrical, musical, and astronomical lore relevant to Plato's writings. In one passage, Theon summarizes various ideas about the distinction between the terms 'one' (ἕν) and 'unit' or 'monad' (μονάς). The passage provides important background to the ideas of the Valentinians and Clement of Alexandria, who assume that their readers are familiar with the notion of the monad's superiority to the one. Although Theon starts off by using the terms 'hen' (ἕν) and 'monad' (μονάς) indiscriminately, he eventually turns to schools of thought that distinguished the terms.[1] The relative obscurity of Theon's passage makes a full translation worth while:[2]

(19.7) καλεῖται δὲ μονὰς ἤτοι ἀπὸ τοῦ μένειν ἄτρεπτος καὶ μὴ ἐξίστασθαι τῆς ἑαυτῆς φύσεως· ὁσάκις γὰρ ἂν ἐφ' ἑαυτὴν πολλαπλασιάσωμεν τὴν μονάδα, μένει μονάς· καὶ γὰρ ἅπαξ ἓν ἕν, καὶ μέχρις ἀπείρου ἐὰν πολλαπλασιάζωμεν τὴν μονάδα, μένει μονάς. ἢ ἀπὸ τοῦ διακεκρίσθαι καὶ μεμονῶσθαι ἀπὸ τοῦ λοιποῦ πλήθους τῶν ἀριθμῶν καλεῖται μονάς.

(19.13) ᾗ δὲ διενήνοχεν ἀριθμὸς καὶ ἀριθμητόν, ταύτῃ καὶ μονὰς καὶ ἕν. ἀριθμὸς μὲν γάρ ἐστι τὸ ἐν νοητοῖς ποσόν, οἷον αὐτὰ ε´ καὶ αὐτὰ ι´, οὐ σώματά τινα οὐδὲ αἰσθητά, ἀλλὰ νοητά· ἀριθμητὸν δὲ τὸ ἐν αἰσθητοῖς ποσόν, ὡς ἵπποι ε´, βόες ε´, ἄνθρωποι ε´. καὶ μονὰς τοίνυν ἐστὶν ἡ τοῦ ἑνὸς ἰδέα ἡ νοητή, ἥ ἐστιν ἄτομος· ἓν δὲ τὸ ἐν αἰσθητοῖς καθ' ἑαυτὸ λεγόμενον, οἷον εἷς ἵππος, εἷς ἄνθρωπος.

---

[1]   18.5 vs. 18.11, 14; 19.6 vs. 19.7.
[2]   Text in angle brackets is excised by the modern editor. Words in square brackets are insertions, by the editor or me, for sense.

(19.21) ὥστ᾿ εἴη ἂν ἀρχὴ τῶν μὲν ἀριθμῶν ἡ μονάς, τῶν δὲ ἀριθμητῶν τὸ ἕν· καὶ τὸ ἓν ὡς ἐν αἰσθητοῖς (20.1) τέμνεσθαί φασιν εἰς ἄπειρον, οὐχ ὡς ἀριθμὸν οὐδὲ ὡς ἀρχὴν ἀριθμοῦ, ἀλλ᾿ ὡς αἰσθητόν. ὥστε ἡ μὲν μονὰς νοητὴ οὖσα ἀδιαίρετος, τὸ δὲ ἓν ὡς αἰσθητὸν εἰς ἄπειρον τμητόν. καὶ τὰ ἀριθμητὰ τῶν ἀριθμῶν εἴη ἂν διαφέροντα τῷ τὰ μὲν σώματα εἶναι, τὰ δὲ ἀσώματα.

(20.5) ἁπλῶς δὲ ἀρχὰς ἀριθμῶν οἱ μὲν ὕστερόν φασι τήν τε μονάδα καὶ τὴν δυάδα, οἱ δὲ ἀπὸ Πυθαγόρου πάσας κατὰ τὸ ἑξῆς τὰς τῶν ὅρων ἐκθέσεις, δι᾿ ὧν ἄρτιοί τε καὶ περιττοὶ νοοῦνται, οἷον τῶν ἐν αἰσθητοῖς τριῶν ἀρχὴν τὴν τριάδα καὶ τῶν ἐν αἰσθητοῖς τεσσάρων πάντων ἀρχὴν τὴν τετράδα καὶ ἐπὶ τῶν ἄλλων ἀριθμῶν κατὰ ταὐτά.

(20.12) οἱ δὲ καὶ αὐτῶν τούτων ἀρχὴν τὴν μονάδα φασὶ καὶ τὸ ἓν πάσης ἀπηλλαγμένον διαφορᾶς ὡς ἐν ἀριθμοῖς, μόνον αὐτὸ ἕν, οὐ τὸ ἕν, τουτέστιν οὐ τόδε τὸ ποιὸν καὶ διαφοράν τινα πρὸς ἕτερον ἓν προσειληφός, ἀλλ᾿ αὐτὸ καθ᾿ αὑτὸ ἕν. οὕτω γὰρ ἂν ἀρχή τε καὶ μέτρον εἴη τῶν ὑφ᾿ ἑαυτὸ ὄντων, καθὸ ἕκαστον τῶν ὄντων ἓν λέγεται, μετασχὸν τῆς πρώτης τοῦ ἑνὸς οὐσίας τε καὶ ἰδέας.

(20.19) Ἀρχύτας δὲ καὶ Φιλόλαος ἀδιαφόρως τὸ ἓν καὶ μονάδα καλοῦσι καὶ τὴν μονάδα ἕν.

(20.20) οἱ δὲ πλεῖστοι προστιθέασι τῷ μονάδα αὐτὴν τὴν πρώτην μονάδα, ὡς οὔσης τινὸς οὐ πρώτης μονάδος, ἥ ἐστι κοινότερον καὶ αὐτὴ μονὰς καὶ ἕν—λέγουσι δὴ καὶ τὸ ἕν[3]—τουτ(21.1)έστιν ἡ πρώτη καὶ νοητὴ οὐσία τοῦ ἑνός, ἑκάστου τῶν πραγμάτων παρέχουσα ἕν· μετοχῇ γὰρ αὐτῆς ἕκαστον ἓν καλεῖται. διὸ καὶ τοὔνομα αὐτοῦ οὐδὲν παρεμφαίνει τί ἓν καὶ τίνος γένους, κατὰ πάντων δὲ κατηγορεῖται, [ὥστε καὶ ἡ μονὰς καὶ ἕν ἐστι,] κἂν τὰ μὲν νοητὰ καὶ παραδείγματα μηδὲν ἀλλήλων διαφέροντα, τὰ δὲ αἰσθητά.

(21.7) ἔνιοι δὲ ἑτέραν διαφορὰν τῆς μονάδος καὶ τοῦ ἑνὸς παρέδοσαν. τὸ μὲν γὰρ ἓν οὔτε κατ᾿ οὐσίαν ἀλλοιοῦται, οὔτε τῇ μονάδι καὶ τοῖς περιττοῖς αἴτιόν ἐστι τοῦ μὴ ἀλλοιοῦσθαι κατ᾿ οὐσίαν, οὔτε κατὰ ποιότητα, αὐτὸ γὰρ μονάς ἐστι καὶ οὐχ ὥσπερ αἱ μονάδες πολλαί, οὔτε κατὰ τὸ ποσόν· οὐδὲ γὰρ συντίθεται ὥσπερ αἱ μονάδες ἄλλῃ μονάδι· ἓν γάρ ἐστι καὶ οὐ πολλά, διὸ καὶ ἑνικῶς καλεῖται ἕν.

(21.14) καὶ γὰρ εἰ παρὰ Πλάτωνι ἑνάδες εἴρηνται ἐν Φιλήβῳ, οὐ παρὰ τὸ ἓν ἐλέχθησαν, ἀλλὰ παρὰ τὴν ἑνάδα, ἥτις ἐστὶ μονὰς μετοχῇ τοῦ ἑνός. κατὰ πάντα δὴ ἀμετάβλητον τὸ ἓν τὸ ὡρισμένον τοῦτο ἐν τῇ μονάδι. ὥστε διαφέροι ἂν τὸ ἓν τῆς μονάδος, ὅτι τὸ μέν ἐστιν ὡρισμένον καὶ πέρας, αἱ δὲ μονάδες ἄπειροι καὶ ἀόριστοι.

---

[3]   Hiller 1878:20: καὶ τὸ ἕν] οὐ τὸ ἕν?

(19.7) It is called 'monad' either from re<u>maining</u> unchangeable and not departing from its own nature.[4] For however often we multiply the monad against itself, it re<u>mains</u> monad. For one once is one, and should we multiply the monad ad infinitum, it remains a monad. Or it is called 'monad' from being distinguished and isolated from the rest of the multitude of numbers.[5]

(19.13) As number differs from numerable thing, so monad differs from one. For number is intelligible quantity, for example, five itself and ten itself, not certain bodies or sense-perceptible objects, but intelligible objects. But a numerable thing is sense-perceptible quantity, for example, five horses, five cows, five people. And so a monad, which is indivisible, is the intelligible form of the one. A sense-perceptible one is spoken of absolutely, for example, "one horse," "one person."

(19.21) Thus, the monad would be the origin of numbers, but the one, of numerable things. And they say that the sense-perceptible one (20) is divided ad infinitum,[6] neither *qua* number nor *qua* origin of number but *qua* sense perceptible. So the monad, being intelligible, is indivisible but the one, as a sense perceptible, is infinitely divisible. And numerable things would differ from numbers in that the former are embodied, and the latter without bodies.

(20.5) Some more recently say simply that the monad and dyad are the origins of numbers, whereas the Pythagoreans [assign this to] the entire subsequent series of limits, through which even and odd are conceived of;[7] for example, the triad is the origin of [all] sense-perceptible threes and the tetrad is the source of every sense-perceptible four, and likewise for the other numbers.

(20.12) Others say that the origin of these very things is the monad and the one removed from every difference that occurs in numbers— only one itself, not the one, that is, not the one exhibiting this quality and certain difference toward another one, but absolute one. So it would be the origin and measure of entities [generated] by itself, by which each entity is called "one," participating in the one's primary substance and form.

(20.19) Archytas and Philolaus call the one 'monad' and the monad 'one,' without differentiation.

---

[4] This sentence is paralleled in Stobaeus *Eclogae* 1.1.8, attributed to Moderatus of Gades (fl. first c. CE). Underlines here highlight his etymology.
[5] This sentence is also paralleled in Stobaeus.
[6] The text from the beginning of the paragraph to this point is paralleled in Stobaeus.
[7] From the previous sentence, "And numerable things … ," to this point is paralleled in Stobaeus.

(20.20) The majority include the primary monad with monad itself, since there is a certain monad that is not primary, but is more commonplace and is monad itself and one—and indeed, they <do not?> call it the one, that is, (21) the primary and intelligible substance of the one, furnishing [the attribute] one to each thing. For each thing is called 'one' by participation in it. Wherefore its name suggests nothing about what is "one" and of what sort, but it is predicated of everything <so that it is both the monad and one>, whether they be intelligible object and paradigms (which do not differ from each other), or sense-perceptible objects.

(21.7) Some hand down a different distinction between the monad and the one. For the one neither changes in substance (nor is it the cause of the monad's and odd numbers' being <un?>alterable in substance), nor [does it change] in quality (for it is a monad, and is not like many monads), nor [does it change] in quantity (for it is not added to another monad, like monads [are]). For it is one and not many, wherefore it is called 'one' in a unifying manner.

(21.14) For even if henads have been mentioned by Plato in *Philebus*, they weren't said [to be] in distinction to the one, but rather in distinction to the henad, which is a monad by virtue of participation in the one. Indeed, in respect to everything in the monad this defined one is unchangeable. So the one would differ from the monad in that the former is defined and limited, whereas monads are infinite and undefined.

Thus Theon, almost certainly following Moderatus, at least in the beginning, explains the etymology of 'monad,' and then offers six opinions concerning the difference between 'monad' and 'hen.'[8] At 19.13, the first of these explanations, he relates the terms to the difference between number (ἀριθμός) and numerable thing (ἀριθμητόν). In distinguishing number from numerables, Theon defines the former as intelligible quantity (literally "quantity in intelligibles"), and not part of the material world (19.15).[9] Numerables, on the other hand, are

---

[8] Theon has been tacitly following Moderatus. Two of the three fragments of Moderatus preserved by Stobaeus have parallels in Theon. Moderatus fragment 1 = Theon 18.3–9 + 19.7–8 + 19.12–13. Moderatus fragment 2 = Theon 19.21–20.1 + 20.4–9. Dodds argues that Theon 19.15 depends on Moderatus (1928:138). Full analysis—indeed a complete corpus of Moderatus' literary fragments—is still needed.

[9] ἀριθμὸς μὲν γάρ ἐστι τὸ ἐν νοητοῖς ποσόν. See also 21.5 and the fragment of Moderatus cited by Simplicius, discussed in chapter 2 above. The distinction between number and numerable thing is maintained by Porphyry, who compares it to the distinction between harmony and something harmonized (*Commentary on Ptolemy's "Harmonics"* 12.2–5).

"quantity in sense-perceptibles," and are predicated of physical objects (19.17).[10] Numerables have bodies, but numbers are bodiless (19.16, 20.5). As numbers are to numerables, Theon/Moderatus claims, so is the monad to the hen (19.13–15). The monad is the intelligible form of the hen, and is indivisible (19.19). Both the monad and the hen are principles: the monad of numbers, and the hen of numerables (19.21–22). The monad and the hen differ, too, in that only the hen may be divided infinitely, because it resides in the corporeal world (19.22–20.4).

Thus we have in this system of thought the notion that the monad stands metaphysically over the hen, with each of the two presiding as the first principle of everything else on its level. The monad presides over objects of intellection; the hen over sense perception.

Theon, again paralleling Moderatus, claims that the next group, more "recent" than the first, identified simply the monad and dyad as the principles of numbers, unlike the Pythagoreans, who claimed that all the intellectual numbers—monad, dyad, triad, tetrad, and so on—provided the principles for the numbers instantiated in the realm of sense perception—*hen*, *duo*, *tria*, *tettares*, and so on (20.5–11). The contrast has echoes in Eudorus, who also contrasts monistic with allegedly later dualist Pythagoreans.[11] But the texts upon which this comparison is based are too short and vague to make the association definite.

A third, unnamed group, according to Theon, claims that the monad and the hen—not just the hen as a quality or point of differentiation, but the absolute hen—were the principle and measure of beings (20.12–19). This absolute one or monad lends its primary substance and form to entities, whereby they can be said to be one. Thus in this monadic system the contrasting terms are 'monad'/'absolute hen' and 'hen.'

Theon presents yet a fourth group, consisting of Archytas and Philolaus, who he says make no distinction between hen and monad (20.19–20 = Archytas, test. 20 = Philolaus, test. 10). Although Syrianus (fl. 430s CE) contradicts him, Theon is probably correct, since Aristotle, one of the more reliable sources for pre-Platonic Pythagoreanism, states that the Pythagoreans called νοῦς both 'monad' and 'hen.'[12] There is no evidence that Plato distinguished the terms, either.[13] Theon, therefore, confirms that the distinction became current only

---

[10] ἀριθμητὸν δὲ τὸ ἐν αἰσθητοῖς ποσόν. See also 21.6.

[11] See chap. 2 above.

[12] Syrianus *Commentary on Aristotle's "Metaphysics"* 151.17–22; Aristotle, fragment 203, in Alexander of Aphrodisias *Commentary on the "Metaphysics"* 39.15.

[13] Note, for instance, the preponderance of 'hen' in the *Parmenides*, and comparative lack of interest in 'monad' as a technical term. The late antique writers who read Plato's writings most closely rely nearly exclusively on 'hen' to describe the metaphysics of arithmetic. Plotinus, for instance, nearly always uses 'hen,' not 'monad,' to describe all his various metaphysical levels of number, in conformity with Plato's *Parmenides*. See Edwards 2006:65–72.

after Plato, and had no currency in classical Pythagoreanism, and that this was still known in his time.

Theon presents a fifth group—the "majority"—that probably overlaps with some of the previous groups. The text is rather muddled, but enough is clear to know that they distinguished one kind of monad from another.[14] They call the lower monad "more common," "monad itself," and "hen" but not "the hen." This is reserved for the higher monad, the chief intelligible essence that furnishes to individual things the property of being one (20.20–21.2). For this group, like the third, something can be said to be "one" by virtue of its participation in the monad (21.2–3). The term 'one' is merely a predicate, an indication of something numerable, whether it be in the intellectual or the sense-perceptible realm. Thus, like the first group, they embrace the hierarchy monad → hen, where the arrow indicates not only metaphysical priority, but a transfer of properties. The system also suggests that the first, absolute monad presides over a realm of intellectual paradigms, which itself presides over the realm of sense perception.

A sixth group distinguishes between 'hen' and 'monad' in a different way. To them, the superiority of the hen to the monads (note the plural) is manifest in its threefold immutability. The hen is immutable in its essence, an immutability that cannot be attributed to the monad or to the odd numbers (21.7–10).[15] Second, the hen is immutable in its quality since it is a monad and is unlike many monads (21.10–11). This vague phrase may refer to the Platonic distinction between ideal numbers, which are unique, and intermediate mathematicals, which resemble each other.[16] Third, the hen is immutable in quantity since it cannot be added the way one monad is combined with another. Otherwise the hen would be many and no longer one (21.11–13). So the monads, treated as countables, change in numerical identity as they mathematically combine. Over the monads resides the hen (occasionally described as a monad). The three immutable aspects of this hen—essence, quality, and quantity—correspond to the first three of Aristotle's categories, pointing to a school of philosophy with strong interest in both Plato and Aristotle.[17] For this group, the ultimate distinction between hen and monad is that the former is defined and is a limit, whereas monads are limitless and indefinite. This arrangement, hen → monad, reverses the schemes found in other groups Theon discusses.

---

[14] See the critical apparatus in Hiller 1878:20–21 for the serious textual problems.

[15] This follows the suggested emendation of Ismaël Bullialdus in the critical apparatus of Hiller's ed. Without this emendation, the parallelism of contrasts in lines 9–14 is broken.

[16] Aristotle *Metaphysics* A6, 987b14–18.

[17] According to my reading of 21.8–13, the punctuation in Hiller's edition should be emended, converting the first comma in line 10 and the comma in line 11 to colons (·) and the colon in line 12 to a comma.

Theon's survey concisely presents the various distinctions made in the second century between 'hen' and 'monad,' and the importance assigned to the subject. To Theon's doxography can be added several other systems from roughly the same period.[18] Alexander Polyhistor, who recounts the Pythagorean doctrine of the generation of numbers, describes the monad's begetting the indefinite dyad, which in turn generates other numbers. Here the monad altogether supplants the traditional hen.[19] Sextus Empiricus too quotes a Pythagorean source that holds to the "first monad" and the "indefinite dyad" as the first principles. The hen derives from this first monad, whereas the number two emerges from the combination of the monad and the indefinite dyad.[20]

Philo uses the terminological distinction to make a theological point about Genesis 24.22.[21] He notes that the monad is to the hen as the archetype is to the copy, and he does so presupposing that this analogy is common knowledge. Although he attests to the doctrine's wide distribution, Philo does not consistently hold to the scheme monad → hen.[22] Clement of Alexandria, seemingly inspired by Philo, embraces the hierarchy hen [θεός] → monad → hen.[23]

Hippolytus reports a rather strange version of the hen → monad doctrine when he claims that the Pythagoreans held to the hierarchy number → monad → n^2 → n^3 (ἀριθμός, μονάς, δύναμις, κύβος).[24] In this arrangement, the hen is associated with the level of "number," and the first monad is the principle of numbers "in their instantiation" (καθ᾽ ὑπόστασιν).[25] All four levels are associa-

---

[18] As well as the reports listed here, see Sextus Empiricus *Against the Physicians* 2.261; pseudo-Pythagoras in pseudo-Justin Martyr (III) *Exhortation to the Nations* 19.2 (ed. Otto 1879:18c); Favonius Eulogius *Disputation on the Dream of Scipio* 3.1–31; John Lydus *On the Months* 2.6; Proclus *Commentary on the "Timaeus"* 1:16.27–29; Boethius *De unitate et de uno*; Asclepius of Tralles *Commentary on Nicomachus of Gerasa* 41.

[19] Alexander Polyhistor, fragment 140 (ed. Müller 1849:240b), in Diogenes Laertius *Lives of the Philosophers* 8.24–25. See also Dillon 1996:127. Alexander agrees with an undatable Pythagorean text ascribed to Xenocrates, who uses 'first monad' in place of 'hen' ("Xenocrates," fragment 120.77, in Sextus Empiricus *Against the Physicians* 2.261–262). For other late antique uses of 'monad' instead of 'hen,' see also idem, 2.282 and Aetius *Placita* 281.5.

[20] Sextus Empiricus *Against the Physicians* 2.276.

[21] Philo *Questions and Answers on Genesis* 4.110.

[22] Philo, at *Who Is the Heir of Divine Things?* 187–190, describes the monad as source of numbers, but does not contrast it to the hen. In *On Rewards and Punishments* 41, he uses 'hen' and 'monad' as a pair, but it is unclear whether he is distinguishing or conflating the terms (cf. idem *On the Unchangeableness of God* 11). At *On the Creation of the World* 98 he uses 'hen' where 'monad' might be expected; at *On Abraham* 122 he uses 'monad' where 'hen' might be called for. At *Allegorical Interpretation* 2.3 he uses both terms together, but specifies that the "one God" (*hen theon*) supersedes the monad. This may be Philo's way of using the language of "one God," native to Judaism, to invert and thereby challenge the monad → hen doctrine so directly stated at *Questions and Answers on Genesis* 4.110. See also p. 127 above.

[23] Clement *Paedagogue* 1.8.71, discussed at p. 126 above.

[24] *Refutation of All Heresies* 1.2.9 = 4.51.7.

[25] *Refutation of All Heresies* 1.2.6 = 4.51.4.

ted with the τετρακτύς, and are thought of as the four parts of the decad. This series also corresponds to that of point, line, plane, and solid. But Hippolytus does not consistently rely upon this scheme. Elsewhere, he slips into language that prioritizes the monad.[26]

Overall, then, late antique authors frequently distinguished between the terms 'monad' and 'hen,' and they took quite different approaches. Although many considered the monad superior to the hen, the variety of opinion shows that there was no consensus, merely a lively interest. Some opinions were an intricate part of an author's overall philosophical commitment. Especially notable among those who most emphasized the distinction was a belief in multiple levels of immaterial reality, particularly levels of mathematicals. The distinction between 'hen' and 'monad' helped to articulate that hierarchy.

[26] See *Refutation of All Heresies* 1.2.2, 6.23.1.

Excursus B

The Pythagorean Symbol of the Τετρακτύς

THE PYTHAGOREANS SYMBOLIZED THE NUMBER TEN with a special term for the first four numbers, the τετρακτύς.[1] The term is probably of Doric origin, but it is unclear exactly how this unusual word, which comes from the root meaning "four," was coined.[2] The term is first attested in texts from the first century,[3] quoting earlier, but not precisely datable, Pythagorean texts. Some may go back to the shadowy origins of early Pythagoreanism, others may be precursors to its reinvention in the late Roman republic. The concept underlying the τετρακτύς has been shown to have predated Pythagoras in non-Greek societies.[4]

The τετρακτύς refers to the first four numbers, which were depicted in the Pythagorean tradition as four rows of ten pebbles arranged in the shape of an isosceles triangle: ∴∵. The figure symbolizes a collection in unity. It also emphasizes that the sum of the first four numbers is ten, a number revered as constituting the foundation of all numbers. By depicting a harmonious arrangement of pebbles, the figure demonstrates the complementary character of arithmetic and geometry, one of the trademarks of the quadrivium. This triangular figure was so well known in the Hellenistic period that Lucian, in one of his satires on the philosophers, has Pythagoras instruct a prospective "buyer" of philosophy to count to four. This four, says Pythagoras, "is ten, and a perfect triangle, and our oath."[5] The oath in question is found in the so-called *Golden*

---

[1] A number of studies have been published on the τετρακτύς. The most extensive are Delatte 1915:249–268 and Kucharski 1952. See also Apatow 1999, Sbordone 1981, Lampropoulou 1975, Burkert 1972 passim, and Haase 1969. On the special use of the term in music see Kárpáti 1993.

[2] See Burkert 1972:222n24 and Delatte 1915:253–254. Cf. Chalcidius *Commentary on the "Timaeus"* 35 (84.9–11), who calls it the *quadratura*.

[3] See n9 below.

[4] See Burkert 1972:474n50.

[5] Lucian *Vitarum auctio* 4: ἃ σὺ δοκέεις τέσσαρα, ταῦτα δέκα ἐστὶ καὶ τρίγωνον ἐντελὲς καὶ ἡμέτερον ὄρκιον. Other explanations of the τετρακτύς as the summation of the first four numbers are found in Aetius *Placita* 1.3.8 (= Diels and Kranz 1957:58b.15); Sextus Empiricus *Against the Logicians* (= *Against the Mathematicians* 6–7) 1.94; and Hippolytus *Refutation of All Heresies* 1.2.8, 4.51.6, 6.23.2–5.

Poem, attributed to Pythagoras and probably the oldest of the Pythagorean texts to mention the τετρακτύς:[6]

οὐ μὰ τὸν ἁμετέρᾳ κεφαλᾷ παραδόντα τετρακτύν,
παγὰν ἀενάου φύσεως ῥιζώματ' ἔχουσαν

No, by the one who grants our head the τετρακτύς,
Fount possessing roots of everlasting nature.[7]

In light of the authentic fragments of Philolaus, these two lines can be reasonably interpreted to suggest that the ancient Pythagoreans held that the first four numbers were forged out of the principles of nature (in Philolaus, these are 'limiters' and 'unlimiteds') to provide a "spring" for the physical world. There is the intriguing possibility that the couplet comes from the same literary milieu as Philolaus' lost work *On Nature*.[8]

---

[6]   On the poem consult Derron 1992. Delatte traces this fragment to Timaeus, of the fourth c. BCE, and an anonymous treatise on arithmology of the second or third c. BCE (1915:249–253). The τετρακτύς is also attested in the ἀκούσματα of the Pythagoreans in Iamblichus *The Pythagorean Way of Life* 82.12 (Diels and Kranz 1957:58c.4). Earlier, Hellenistic Pythagorean texts that mention the τετρακτύς are "Lysis," fragment 4.4 (= Diels and Kranz 1957:46.4), in Athenagoras *Legatio* 6.1, and an anonymous philosopher paraphrased in Photius *Bibliotheca* 439a7–8 (Bekker 1825). Thesleff 1961 tentatively dates these to the fourth and third c. BCE, respectively. Also see Philolaus, fragment 11 (found in Lucian *De lapsu in salutandum* 5), of dubious date and authenticity.

[7]   Sextus Empiricus *Against the Logicians* (= *Against the Mathematicians* 6–7) 1.94. The two lines are reproduced with significant differences in other authors: pseudo-Pythagoras *Golden Poem* 47–48; Aetius *Placita* 282.3–7; Nicomachus of Gerasa, in *Theology of Arithmetic* 22.21–22; Sextus Empiricus *Against the Mathematicians* 4.2; Theon of Smyrna *Mathematics Useful for Reading Plato* 94.6–7; Hippolytus *Refutation of All Heresies* 6.23.4; Porphyry *Life of Pythagoras* 20.18–19; Iamblichus *The Pythagorean Way of Life* 29.162.17–18; Julian *To the Untaught Dogs* 15.34; Stobaeus *Eclogae* 1.10.12.72–73; Hierocles *On the "Golden Poem"* 20; Damascius *Commentary on the "Parmenides"* 63.29; Proclus *Commentary on the "Timaeus"* 2.53.6. For analysis of these differences, see Delatte 1915:249–253. Possibly even Xenocrates (frags. 101–102, Isnarde Parente 1982), when he suggests that "the universe consists of the One and the Everlasting" (συνεστάναι τὸ πᾶν ἐκ τοῦ ἑνὸς καὶ τοῦ ἀενάου), uses the Pythagorean τετρακτύς as a symbol of matter. Such an ancient testimony does not help date the *Golden Poem*, but it does establish the antiquity of the motif. See Dillon 1996:24.

[8]   The argument in outline is this: Philolaus is concerned with "nature," an important concept in the couplet. One may interpret the second line to say that the τετρακτύς is the root of eternal nature. But it is equally possible to read the genitives so that the roots producing the τετρακτύς derive from eternal nature. In this case, number is subordinate to and derived from eternal principles such as unlimiteds and limiters, which as Huffman has stressed is the nature of Philolaus' philosophy (1993). Further, the epithets for the τετρακτύς in the ἀκούσματα—the oracle at Delphi, harmony, and the location of the Sirens (Iamblichus *The Pythagorean Way of Life* 82.12 [= DK 58c.4])—are ancient, nonmathematical, and (with the exception of 'harmony') nonphilosophical. Does the couplet, then, derive from the μαθηματικοί faction of ancient Pythagoreanism (see Burkert 1972)? If so, the question of origins remains: did Philolaus write the couplet, was the couplet composed in light of Philolaus' book, did the couplet come from the older Pythagorean oral tradition, or are the couplet and Philolaus independent of each other but dependent upon a common tradition?

In the first and the second centuries, probably as a result of the revival and reinvention of Pythagoreanism, the τετρακτύς entered non-Pythagorean literary circles as a metaphor.[9] Because legend had it that the Pythagorean tradition was a secret one, and because the τετρακτύς was seen as the basis of their oath, the symbol took on special mystical significance that extended beyond its primary mathematical meaning. Like other Pythagorean symbols, it could connect disparate foursomes in the world. Theon of Smyrna collected eleven different quaternities found in the world, calling them τετρακτύες.[10] His examples range from the mathematical (point, line, plane, and solid) to anthropological (ages of human beings: child, teenager, adult, and elder). Another author, of unknown date, drew up a similar list of six τετρακτύες, three of which have no parallels in Theon or other ancient authors.[11] This reflects the popular, literary character of Pythagoreanism. An author could theoretically take any foursome, relate it to the τετρακτύς, and thereby tap into the world of Pythagorean symbolism.

Similarly, an author could postulate a foursome and describe its internal relations so as to invoke Pythagorean overtones. In return, Pythagorean imagery and terminology used to structure a sequence of four elements could reinforce and supplement the lore of the τετρακτύς. In philosophy and theology in late antiquity this phenomenon is common. The internal structure of a philosophical or theological quaternity generally follows one of two patterns. In the first, the initial element of the quaternity begets the second, the second begets the third, and the third the fourth—a linear progression. Examples of this kind of quaternity are the number series one, two, three, and four; or the geometric series point, line, plane, and solid.[12] In the second pattern, the author conceives of the foursome as two complementary, hierarchical pairs. The pairs can be expressed in the relation $A : B :: C : D$, like the four corners of a rectangle. Examples of this are Neoplatonic theories of epistemology or the quadrivium.

Polemic aside, the accusations of Irenaeus, Hippolytus, and other apologists who charge the Valentinians and others of teaching the Pythagorean τετρακτύς in the guise of Christian doctrine cannot be dismissed as fraudulent. The apologists' rhetoric is often excessive, as expected in the genre. But they

---

[9]   Search results from the TLG (E) are instructive on the popularity of the term. Discounting the statistics for the Hellenistic Pythagorean texts (very difficult to date), τετρακτύς appears in no texts BCE, five times in the first century CE, forty-six in the second, twenty-six in the third, twenty-four in the fourth, and twenty-four in the fifth.

[10] *Mathematics Useful for Reading Plato* 94–99.

[11] This very brief treatise, called Τετρακτὺν τὴν τὰ πάντα διατείνουσαν καὶ διοιροῦσαν τετραχῆ πάντα (*The Tetraktys Suspending and Apportioning All Things Four-fold*), is found in Paris gr. 1185 suppl., fol. 62v., and is published in Delatte 1915:187.

[12] Cf. the examples provided by Alexander Polyhistor and Hippolytus, discussed at pp. 181–182 above.

correctly recognize that certain doctrines—for example the Monotes-Henotes-Monas-Hen doctrine assigned to an unnamed Valentinian and Marcus—are attuned to a Pythagorean model of foursomes.[13]

The use of τετρακτύς in Christian literature reflects the early but transient suspicion the orthodox had of gnostic opponents. In the second and third century, orthodox Christian authors use τετρακτύς in a disparaging or neutral way.[14] But in the fourth century, after Valentinianism waned, Christians freely used it to symbolize Christian truths, such as the unity of the four Gospels and the fourfold character of Christian virtue.[15]

---

[13] See pp. 31–33 and 75 above.

[14] The most favorable religious use is by Athenagoras *Legatio* 6.1, who simply presents it as a part of the philosophical apparatus that undermines polytheism. Other instances are found at Irenaeus *Against Heresies* 1.1.1, 1.1.13, 1.8.4–5, 1.8.10, 1.11.1–2; Clement of Alexandria *Stromateis* 2.23.138.6; Hippolytus *Refutation of All Heresies* 1.2.9; 4.51.7; 6.23.4–5; 6.24.1; 6.34.1; 6.44.1; 6.45.2; and Anatolius of Laodicea *On the Decad* 5.11, 8.1, 15.20.

[15] Eusebius of Caesarea *Church History* 3.25.1; Theodoret of Cyrus *Letters* 131.112, 146.200; Evagrius of Pontus *On Prayer* pref. (PG 79.1165d).

Excursus C

The Dyadic Character of
A Valentinian Exposition

T HE NAG HAMMADI TEXT *A VALENTINIAN EXPOSITION* has been said to champion monadic rather than dyadic Valentinianism. Four reasons are generally given: First, the Father is described as being alone, and is called 'Monad' (NH 11.2:22.19–23.21). Second, Silence, the usual consort of the Father in Irenaeus' reports, comes on the scene slowly, through synonyms such as 'quietness' (ΠΚΑΡΩϤ) and 'tranquility' (ΠϹϬΡΑϩⲦ), as if to downplay any notion that Silence is coeval with or consort to the Father, and to make Silence an abstraction, merely the tranquility of the Father's solitude. Third, it seems that in *A Valentinian Exposition* the "Uncreated One"—understood to be Only Begotten, the third aeon—generates the second Tetrad on his own, thus imitating the primal solitude of the Father (NH 11.2:29.29–30).[1] That is, by generating without a consort, Only Begotten reveals that the Father is also without consort. Finally, *A Valentinian Exposition*'s epithet for the Father, "root of all," the familiar Valentinian designation for the primary Tetrad or the ensuing Ogdoad (NH 11.2:22.20, 33–34; 23.19), seems to enhance the monadic status of the Father. These arguments motivate John Turner and Elaine Pagels to classify the text as monadic Valentinian.[2] The arguments are not persuasive, and they do not reckon with stronger evidence for the dyadic character of the text.

First, although the Father seems to be called 'Monad,' he is also said to exist *in* the Monad (ϩⲚ ⲦⲘⲞⲚⲀϹ), and even to exist *in* the Dyad and pair ([ϩⲚ ⲦⲀ] ⲨⲀϹⲀⲨⲰϩⲚ ΠϹⲀⲈⲒⲰ; NH 11.2:22.21, 26). Later on, an unidentified subject, presumably the Father, is said to exist *in* the Monad, Dyad, and Tetrad (NH 11.2:25.19–20). How can an entity be in something, yet be that something as well? How can such a solitary entity dwell in the Dyad or Tetrad?

---

[1] 'Uncreated One' is probably not the likely name. See Thomassen 2006:232n53.
[2] Turner 1990:96–99, 160–161 (at 29.25–30, 29.29–35).

The paradox may be explained by the two passages that seem to declare that the Father is the Monad. A close examination shows that, just like *The Tripartite Tractate*, *A Valentinian Exposition* creates a complex synthesis of monism and dualism. The first passage, based upon Turner's reconstruction [ⲚⲈϤϢⲞⲞ] Ⲡ ̄ⲘⲘⲞⲚⲀⲤ, more likely means "[he existed] monadically," than "[he was] the Monad" (NH 11.2:22.24).[3] The verb Turner supplies depends not on the manuscript—the fragmentary blip taken to be Ⲡ leaves much to be desired—but on analogy with the second passage, where some unspecified subject is said to be "[the] Root [of the All] and Monad (ⲀⲨⲰⲘⲞⲚⲀⳓ [ⲠⲈ]) without any[one] before him" (NH 11.2:23.19–21).[4] But here the lack of the definite article before ⲘⲞⲚⲀⳓ suggests that the subject is not the Monad but *a* monad, i.e. something acting *qua* unit. Elsewhere in *A Valentinian Exposition*, important reified entities such as the Monad are always identified with the definite article. Further, although the Father has no one who exists before him, this is not the same as saying "[He dwells alone]."[5] So the notion that the Father exists equally in the Monad, Dyad, and Tetrad I take to mean that the Father exists in several ways—monadically, dyadically, and tetradically. The text is asserting not so much what the Father is, but his modes of existence and action.

Silence (ⲦⲤⲒⲄⲎ) only appears to take the stage slowly, on page 22 (the second argument for classifying *A Valentinian Exposition* as monadic). Nearly the entire upper half of the page, the beginning of the treatise, is missing. This missing text is the proper basis for determining the status of Silence and whether or not she is introduced at the outset. On page 22, no relationship is explicitly established among Silence, Quietness (line 22), and Tranquility (line 23), so it is impossible to say whether the author means the latter two terms to delay the introduction of the former, as Turner suggests.[6] If ⲦⲤⲒⲄⲎ was introduced in the upper part of the folio, the later occurrence of ⲠⲔⲀⲢⲰϤ and ⲠⲤⲞⲢⲀϩ̄Ⲧ̄ would only amplify, not attenuate, Silence's role as consort of the Father. Even if Silence was not introduced at the top of the folio, the order of the extant text mirrors the presentation of Irenaeus' extended Valentinian system, which was dyadic. In that system, Depth, the first aeon, "abides in great rest and peace" (ἐν ἡσυχίᾳ καὶ ἠρεμίᾳ πολλῇ γεγονέναι). Silence is then introduced in the next sentence. In Irenaeus' report, the delay in introducing Silence does not diminish the system's dyadic character. Indeed, the order ἡσυχία, ἠρεμία, Σιγή mirrors exactly ⲠⲔⲀⲢⲰϤ, ⲠⲤⲞⲢⲀϩ̄Ⲧ̄, ⲦⲤⲒⲄⲎ.

---

[3] Crum 1962:578–579. I thank Janet Timbie for her suggestions, here and throughout this section.

[4] Translations of this text are Turner's. For the broken letter see Robinson 1973:28.

[5] NH 11.2:22.24–25, 38; 23.20–21; 22.22. See Turner 1990:97. In Robinson 1973:28, there is no apparent survival of what Turner indicates to be ϥ, so the entire conjecture, [ⲈϤϢⲞⲞⲠⲞⲨⲀⲈⲈⲦ] ϥ, depends upon the editor's conjecture that a monadic system is at work.

[6] Turner 1990:97.

There is evidence that Silence plays the same important role in *A Valentinian Exposition* that she does in Irenaeus' first Valentinian system. She forms with the Ineffable the primal Dyad, and is second to him (NH 11.2:22.26; 29.31–33; 23.21–22), thereby suggesting a dyadic more than a monadic model. Also, the will of the Father, according to *A Valentinian Exposition*, is to allow nothing to happen in the Pleroma without a syzygy. This would be strange counsel if the Father himself were not the archetype (NH 11.2:36.28–31).

According to Turner's edition, "[the Uncreated One] projected Word and Life," thus crediting Only Begotten—the third member of the primal Tetrad—with the generation of the first syzygy of the second Tetrad (NH 11.2:29.29–30). The reconstructed text seems to suggest that Only Begotten creates the syzygy on his own, just as the Father dwells in monadic solitude, although neither Turner nor Pagels is completely clear on this matter.[7] This reconstruction, however, contradicts other parts of *A Valentinian Exposition*, as well as Irenaeus' extended system, which in so many other respects harmonizes well. For both Irenaeus' Valentinians and *A Valentinian Exposition*, the first Tetrad, not Only Begotten alone, projects the second Tetrad (NH 11.2:29.25–26, 35–37; Irenaeus, *Against Heresies* 1.11.1; see n7). In contrast, the monadic Valentinianism of Hippolytus has no Tetrad in its upper emanations, since they are grouped only in pairs, not Tetrads. Indeed, *A Valentinian Exposition* declares against the monadic Valentinians of Hippolytus that there are a total of thirty emanations, not twenty-eight, before Wisdom's fall. Turner's reconstruction of 29.29 is debatable; other readings consistent on every level with the text's meaning and grammar can be supplied so as to place *A Valentinian Exposition* on the dyadic side of Valentinian thought.[8]

---

[7] Turner suggests that the "non-creature" could be the syzygy Only Begotten and Truth (1990:161), but this seems to render pointless his distinction between dyadic and monadic Valentinian accounts of the generation of the second Tetrad (1990:160). After all, the parallels Turner presents differ only as to whether the entire primal Tetrad, or merely the syzygy Only Begotten–Father–Mind/Truth, projects the second Tetrad. But this distinction has nothing to do with whether the system is monadic or dyadic. Even if *A Valentinian Exposition* says Only Begotten has alone begotten the second Tetrad, this conforms more closely to the dyadic Valentinian account at Irenaeus *Against Heresies* 1.1.1 (which has two Tetrads) than it does with the monadic one at Hippolytus *Refutation of All Heresies* 6.29.6–7 (which has no Tetrads: the upper level of the Pleroma has six entities).

[8] Turner reads [c]ιⲁ [ⲡⲁⲧⲥⲱ]ⲱⲛⲧ [ⲛ̅ⲁⲉ ⲁ]ϥ̅ⲧⲉⲩⲟ. In Robinson 1973 only the ⲛ is clear in the second word. The stroke interpreted as ⲱ appears too far below the baseline to be an omega. Cf. the omegas at lines 32, 33. There may be several ways to restore the middle of the line; I might suggest one, [ϩⲛ̅ ⲧⲙⲁϩ]ⲥ̅ⲛ̅ⲧ[ⲉ ⲛ̅ⲁⲉ ⲁ]ϥ̅ⲧⲉⲩⲟ ("secondarily he projected" or "in the second he projected"). This option originates from the observation that the "Second" has already been reified as a modality on p. 23. There, the unspecified subject (Turner, postulates the Father or Root of All [1990:154]) does various things on three different levels: coming forth in the realm of the 360th; revealing his will in the Second; and spreading himself in the Fourth (23.26–31). This so-called Second may be Silence herself (cf. 22.26–27), or it may be the second syzygy, which dwells

The fourth argument for the monadic theology of *A Valentinian Exposition* is based on the observation that the epithet "Root of All" is applied only to the Father. This is unconvincing on its own. Given that Irenaeus' Valentinian (and dyadic) system calls Νοῦς "source of all" and the Forefather "root without source," this may be yet further evidence for a modified dyadic system at the heart of *A Valentinian Exposition.*[9] An epithet is meant to summarize, not explain, the status of its subject. In *A Valentinian Exposition* the title "root of all" is never explained in securely read text, and therefore I believe it unwise to use it as a determining factor.

Based on all the evidence above, it seems to me that *A Valentinian Exposition* falls on the dyadic side of Valentinianism, even though it attempts to preserve a unique, monadic quality in the Father, as many dyadic Valentinian systems do.

---

in, and originates from, Silence (23.21–22). In Irenaeus' report, Only Begotten, the male part of the second syzygy, projects the third, Word and Life (*Against Heresies* 1.1.1). In the interests of brevity, Irenaeus may have omitted any mention of Truth's participation; thus the original idea would have been that the entire second syzygy projects the second Tetrad. This notion is echoed by Hippolytus *Refutation of All Heresies* 6.29.6–7. Thus, in my reconstruction, an unspecified subject (the entire primal Tetrad? the Father alone?) projects Word and Life in a second phase of emanations, or by means of the Second—again this could be Silence or the second syzygy. This suggestion presumes that the top half of fol. 29 specified the context and meaning of "Second." Something should happen "first," such as what is specified at 25.20–21, where something—apparently the Father—"first brings forth" Only Begotten and Limit. This reconstruction provides a meaning quite consonant with dyadic Valentinianism. I mean to suggest not that this is the only way to reconstruct the text, but that we need not let our presumption that *A Valentinian Exposition* comes from monadic Valentinianism—a presumption built upon a false dichotomy—determine the restoration of the text. On the complexities of the "Second" in *A Valentinian Exposition*, see Turner 1990:155–156.

9 Νοῦς: Irenaeus *Against Heresies* 1.1.1, ἀρχὴν τῶν πάντων. Forefather: *Against Heresies* 1.2.1, τὴν ἄναρχον ῥίζαν.

Appendix
Greek Texts

Irenaeus *Revelation to Marcus*

Chapter 4 above (pp. 62–80).

(1.14.1) Οὗτος <οὖν ὁ> Μάρκος μήτραν καὶ ἐκδοχεῖον τῆς Κολαρβάσου Σιγῆς αὐτὸν μονώτατον γεγονέναι λέγων, ἅτε Μονογενὴς ὑπάρχων, [αὐτὸ] τὸ σπέρμα τὸ κατατεθὲν εἰς αὐτὸν ὧδέ πως ἀπεκύησεν. Αὐτὴν τὴν πανυπερτάτην ἀπὸ τῶν ἀοράτων καὶ ἀκατονομάστων τόπων Τετράδα κατεληλυθέναι σχήματι γυναικείῳ πρὸς αὐτόν, ἐπειδή, φησί, τὸ ἄρρεν αὐτῆς ὁ κόσμος φέρειν οὐκ ἠδύνατο, καὶ μηνῦσαι αὐτήν, τίς ἦν, καὶ τὴν τῶν πάντων γένεσιν, ἣν οὐδενὶ πώποτε οὔτε θεῶν οὔτε ἀνθρώπων ἀπεκάλυψε, τούτῳ μονωτάτῳ διηγήσασθαι, οὕτως εἰποῦσαν· Ὅτε τὸ πρῶτον ὁ Πατὴρ <οὗ Πατὴρ> οὐδείς, ὁ ἀνεννόητος καὶ ἀνούσιος, ὁ μήτε ἄρρεν μήτε θῆλυ, ἠθέλησεν αὐτοῦ τὸ ἄρρητον ῥητὸν γενέσθαι καὶ τὸ ἀόρατον μορφωθῆναι, ἤνοιξε τὸ στόμα καὶ προήκατο Λόγον ὅμοιον αὐτῷ· ὃς παραστὰς ἐπέδειξεν αὐτῷ ὃ ἦν, αὐτὸς τοῦ ἀοράτου μορφὴ φανείς. Ἡ δὲ ἐκφώνησις τοῦ ὀνόματος ἐγένετο τοιαύτη· ἐλάλησε λόγον τὸν πρῶτον τοῦ ὀνόματος αὐτοῦ, ἥτις ἦν ἀρχή, καὶ ἦν ἡ συλλαβὴ αὐτοῦ στοιχείων τεσσάρων· ἐπισυνῆψεν τὴν δευτέραν, καὶ ἦν καὶ αὐτὴ στοιχείων τεσσάρων· ἑξῆς ἐλάλησε τὴν τρίτην, καὶ ἦν αὐτὴ στοιχείων δέκα· καὶ τὴν μετὰ ταῦτα ἐλάλησε, καὶ ἦν [καὶ] αὐτὴ στοιχείων δεκαδύο. Ἐγένετο οὖν ἡ ἐκφώνησις τοῦ ὅλου ὀνόματος στοιχείων μὲν τριάκοντα, συλλαβῶν δὲ τεσσάρων. Ἕκαστον δὲ τῶν στοιχείων ἴδια γράμματα καὶ ἴδιον χαρακτῆρα καὶ ἰδίαν ἐκφώνησιν καὶ σχήματα καὶ εἰκόνας ἔχειν· καὶ μηδὲν αὐτῶν εἶναι ὃ τὴν ἐκείνου καθορᾷ μορφὴν οὗπερ αὐτὸ στοιχεῖόν ἐστιν, ἀλλὰ οὐδὲ γινώσκειν αὐτό· οὐδὲ μὴν τὴν τοῦ πλησίου αὐτοῦ ἕκαστον ἐκφώνησιν γινώσκειν, ἀλλὰ ὃ αὐτὸ ἐκφωνεῖ, ὡς τὸ πᾶν ἐκφωνοῦν, τὸ ὅλον ἡγεῖσθαι ὀνομάζειν. Ἕκαστον γὰρ αὐτῶν, μέρος ὄν τοῦ ὅλου, τὸν ἴδιον ἦχον ὡς τὸ πᾶν ὀνομάζειν, καὶ μὴ παύσασθαι ἠχοῦντα, μέχρις ὅτου ἐπὶ τὸ ἔσχατον γράμμα τοῦ ἐσχάτου στοιχείου μονογλωσσήσαντα καταντῆσαι. Τότε δὲ καὶ τὴν ἀποκατάστασιν τῶν ὅλων ἔφη γενέσθαι, ὅταν τὰ πάντα κατελθόντα εἰς τὸ ἓν γράμμα μίαν καὶ τὴν αὐτὴν ἐκφώνησιν ἠχήσῃ· ἧς ἐκφωνήσεως εἰκόνα τὸ ἀμὴν ὁμοῦ λεγόντων ἡμῶν ὑπέθετο εἶναι. Τοὺς δὲ φθόγγους ὑπάρχειν τοὺς μορφοῦντας τὸν ἀνούσιον καὶ ἀγέννητον Αἰῶνα· καὶ εἶναι τούτους μορφάς, ἃς ὁ Κύριος Ἀγγέλους εἴρηκε, τὰς διηνεκῶς βλεπούσας τὸ πρόσωπον τοῦ Πατρός.

(1.14.2) Τὰ δὲ ὀνόματα τῶν στοιχείων τὰ κοινὰ καὶ ῥητὰ Αἰῶνας καὶ λόγους καὶ ῥίζας καὶ σπέρματα καὶ πληρώματα καὶ καρποὺς ὠνόμασε· τὰ δὲ καθ' ἕνα αὐτῶν καὶ ἑκάστου ἴδια ἐν τῷ ὀνόματι τῆς Ἐκκλησίας ἐμπεριεχόμενα νοεῖσθαι ἔφη. Ὧν στοιχείων τοῦ ἐσχάτου στοιχείου τὸ ὕστερον γράμμα φωνὴν προήκατο τὴν ἑαυτοῦ, οὗ ὁ ἦχος ἐξελθὼν κατ' εἰκόνα τῶν στοιχείων στοιχεῖα ἴδια ἐγέννησεν, ἐξ ὧν τά τε ἐνταῦθα διακεκοσμῆσθαί φησι καὶ τὰ πρὸ τούτων γεγενῆσθαι. Τὸ μέντοι γράμμα αὐτό, οὗ ὁ ἦχος ἦν συνεπακολουθῶν τῷ ἤχῳ κάτω, ὑπὸ τῆς συλλαβῆς τῆς ἑαυτοῦ ἀνειλῆφθαι ἄνω λέγει εἰς ἀναπλήρωσιν τοῦ ὅλου, μεμενηκέναι δὲ εἰς τὰ κάτω τὸν ἦχον ὥσπερ ἔξω ῥιφέντα. Τὸ δὲ στοιχεῖον αὐτό, ἀφ' οὗ τὸ γράμμα σὺν τῇ ἐκφωνήσει τῇ ἑαυτοῦ συγκατῆλθε κάτω, γραμμάτων εἶναί φησι τριάκοντα, καὶ ἓν ἕκαστον τῶν τριάκοντα γραμμάτων ἐν ἑαυτῷ ἔχειν ἕτερα γράμματα, δι' ὧν τὸ ὄνομα τοῦ γράμματος ὀνομάζεται, καὶ αὖ πάλιν τὰ ἕτερα δι' ἄλλων ὀνομάζεσθαι γραμμάτων, καὶ τὰ ἄλλα δι' ἄλλων, ὡς εἰς ἄπειρον ἐκπίπτειν τὸ πλῆθος τῶν γραμμάτων. Οὕτω δ' ἂν σαφέστερον μάθοις τὸ λεγόμενον· τὸ δέλτα στοιχεῖον γράμματα ἐν ἑαυτῷ ἔχει πέντε, αὐτό τε τὸ δέλτα καὶ τὸ ε καὶ τὸ λάμβδα καὶ τὸ ταῦ καὶ τὸ ἄλφα, καὶ ταῦτα πάλιν τὰ γράμματα δι' ἄλλων γράφεται γραμμάτων, καὶ τὰ ἄλλα δι' ἄλλων. Εἰ οὖν ἡ πᾶσα ὑπόστασις τοῦ δέλτα εἰς ἄπειρον ἐκπίπτει, ἀεὶ ἄλλων ἄλλα γράμματα γεννώντων καὶ διαδεχομένων ἄλληλα, πόσῳ μᾶλλον ἐκείνου τοῦ στοιχείου μεῖζον εἶναι τὸ πέλαγος τῶν γραμμάτων; Καὶ εἰ τὸ ἓν γράμμα οὕτως ἄπειρον, ὅρα ὅλου τοῦ ὀνόματος τὸν βυθὸν τῶν γραμμάτων, ἐξ ὧν τὸν Προπάτορα ἡ Μάρκου Σιγὴ συνεστάναι ἐδογμάτισε. Διὸ καὶ τὸν Πατέρα ἐπιστάμενον τὸ ἀχώρητον αὐτοῦ δεδωκέναι τοῖς στοιχείοις, ἃ καὶ Αἰῶνας καλεῖ, ἑνὶ ἑκάστῳ αὐτῶν τὴν ἰδίαν ἐκφώνησιν ἐκβοᾶν, διὰ τὸ μὴ δύνασθαι ἕνα τὸ ὅλον ἐκφωνεῖν.

(1.14.3) Ταῦτα δὲ σαφηνίσασαν αὐτῷ τὴν Τετρακτὺν εἰπεῖν· Θέλω δέ σοι καὶ αὐτὴν ἐπιδεῖξαι τὴν Ἀλήθειαν· κατήγαγον γὰρ αὐτὴν ἐκ τῶν ὕπερθεν δωμάτων, ἵν' εἰσίδῃς αὐτὴν γυμνὴν καὶ καταμάθῃς τὸ κάλλος αὐτῆς, ἀλλὰ καὶ ἀκούσῃς αὐτῆς λαλούσης καὶ θαυμάσῃς τὸ φρόνιμον αὐτῆς. Ὅρα οὖν κεφαλὴν <αὐτῆς> ἄνω τὸ α καὶ τὸ ω, τράχηλον δὲ β καὶ ψ, ὤμους ἅμα χερσὶν γ καὶ χ, στήθη δ καὶ φ, διάφραγμα ε καὶ υ, κοιλίαν ζ καὶ τ, αἰδοῖα η καὶ σ, μηροὺς θ καὶ ρ, γόνατα ι καὶ π, κνήμας κ καὶ ο, σφυρὰ λ καὶ ξ, πόδας μ καὶ ν. Τοῦτό ἐστι τὸ σῶμα τῆς κατὰ τὸν μάγον Ἀληθείας, τοῦτο τὸ σχῆμα τοῦ στοιχείου, οὗτος ὁ χαρακτὴρ τοῦ γράμματος. Καὶ καλεῖ τὸ στοιχεῖον τοῦτο Ἄνθρωπον· εἶναί τε πηγήν φησιν αὐτὸ παντὸς λόγου καὶ ἀρχὴν πάσης φωνῆς καὶ παντὸς ἀρρήτου ῥῆσιν καὶ τῆς σιωπωμένης Σιγῆς στόμα. Καὶ τοῦτο μὲν τὸ σῶμα αὐτῆς· σὺ δὲ μετάρσιον ἐγείρας τὸ τῆς διανοίας νόημα, τὸν αὐτογεννήτορα καὶ πατροδότορα Λόγον ἀπὸ στομάτων Ἀληθείας ἄκουε.

(1.14.4) Ταῦτα δὲ ταύτης εἰπούσης, προσβλέψασαν αὐτῷ τὴν Ἀλήθειαν καὶ ἀνοίξασαν τὸ στόμα λαλῆσαι λόγον, τὸν δὲ λόγον ὄνομα γενέσθαι, καὶ τὸ ὄνομα τοῦτο εἶναι ὃ γινώσκομεν καὶ λαλοῦμεν, Χρ{ε}ιστὸν Ἰησοῦν, ὃ καὶ ὀνομάσασαν αὐτὴν παραυτίκα σιωπῆσαι. Προσδοκῶντος δὲ τοῦ Μάρκου πλεῖόν τι μέλλειν αὐτὴν λέγειν, πάλιν ἡ Τετρακτὴς παρελθοῦσα εἰς τὸ μέσον φησίν· Ὡς εὐκαταφρόνητον ἡγήσω τὸν λόγον, ὃν ἀπὸ στομάτων τῆς Ἀληθείας ἤκουσας; Οὐ τοῦθ' ὅπερ οἶδας καὶ δοκεῖς ἔχειν παλαιόν ἐστιν ὄνομα· φωνὴν γὰρ μόνον ἔχεις αὐτοῦ, τὴν δὲ δύναμιν ἀγνοεῖς. Ἰησοῦς μὲν γάρ ἐστιν ἐπίσημον ὄνομα, ἓξ ἔχον γράμματα, ὑπὸ πάντων τῶν τῆς κλήσεως γινωσκόμενον· τὸ δὲ παρὰ τοῖς Αἰῶσι τοῦ Πληρώματος, πολυμερὲς τυγχάνον, ἄλλης ἐστὶν μορφῆς καὶ ἑτέρου τύπου, γινωσκόμενον ὑπ' ἐκείνων τῶν συγγενῶν, ὧν τὰ Μεγέθη παρ' αὐτῷ ἐστι διὰ παντός.

(1.14.5) Ταῦτ' οὖν τὰ παρ' ἡμῖν εἰκοσιτέσσαρα γράμματα ἀπορροίας ὑπάρχειν γίνωσκε τῶν τριῶν Δυνάμεων εἰκονικὰς τῶν περιεχουσῶν τὸν ὅλον τῶν ἄνω στοιχείων ἀριθμόν. Τὰ μὲν γὰρ ἄφωνα γράμματα ἐννέα νόμισον εἶναι τοῦ Πατρὸς καὶ τῆς Ἀληθείας, διὰ τὸ ἀφώνους αὐτοὺς εἶναι, τουτέστιν ἀρρήτους καὶ ἀνεκλαλήτους· τὰ δὲ ἡμίφωνα, ὀκτὼ ὄντα, τοῦ Λόγου καὶ τῆς Ζωῆς, διὰ τὸ μέσα ὥσπερ ὑπάρχειν τῶν τε ἀφώνων καὶ τῶν φωνηέντων καὶ ἀναδέχεσθαι τῶν μὲν ὕπερθεν τὴν ἀπόρροιαν, τῶν δ' ὑπ' αὐτὰ τὴν ἀναφοράν· τὰ δὲ φωνήεντα, καὶ αὐτὰ ἑπτὰ ὄντα, τοῦ Ἀνθρώπου καὶ τῆς Ἐκκλησίας, ἐπεὶ διὰ τοῦ Ἀνθρώπου ἡ φωνὴ προελθοῦσα ἐμόρφωσε τὰ ὅλα· ὁ γὰρ ἦχος τῆς φωνῆς μορφὴν αὐτοῖς περιεποίησεν. Ἔστιν οὖν ὁ μὲν Λόγος ἔχων καὶ ἡ Ζωὴ τὰ ὀκτώ, ὁ δὲ Ἄνθρωπος καὶ ἡ Ἐκκλησία τὰ ἑπτά, ὁ δὲ Πατὴρ καὶ ἡ Ἀλήθεια τὰ ἐννέα. Ἐπὶ δὲ τοῦ ὑστερήσαντος λόγου ὁ ἀφεδρασθεὶς ἐν τῷ Πατρὶ κατῆλθεν, ἐκπεμφθεὶς ἐπὶ τὸν ἀφ' οὗ ἐχωρίσθη, ἐπὶ διορθώσει τῶν πραχθέντων, ἵνα ἡ τῶν Πληρωμάτων ἑνότης ἰσότητα ἔχουσα καρποφορῇ μίαν ἐν πᾶσι τὴν ἐκ πάντων Δύναμιν. Καὶ οὕτως ὁ τῶν ἑπτὰ τὴν τῶν ὀκτὼ ἐκομίσατο δύναμιν καὶ ἐγένοντο οἱ τρεῖς τόποι ὅμοιοι τοῖς ἀριθμοῖς, Ὀγδοάδες ὄντες, οἵτινες τρὶς ἐφ' ἑαυτοὺς ἐλθόντες τὸν τῶν εἰκοσιτεσσάρων ἀνέδειξαν ἀριθμόν. Τὰ μέντοι τρία στοιχεῖα, ἅφησιν αὐτὸς τῶν τριῶν ἐν συζυγίᾳ Δυνάμεων ὑπάρχειν, ἅ ἐστιν ἕξ, ἀφ' ὧν ἀπερρύη τὰ εἰκοσιτέσσαρα στοιχεῖα, τετραπλασιασθέντα τῷ τῆς ἀρρήτου Τετράδος λόγῳ, τὸν αὐτὸν αὐτοῖς ἀριθμὸν ποιεῖ, ἅπερ φησὶ τοῦ ἀνονομάστου ὑπάρχειν. Φορεῖσθαι δὲ αὐτὰ ὑπὸ τῶν τριῶν Δυνάμεων, εἰς ὁμοιότητα τοῦ ἀοράτου, ὧν στοιχείων εἰκόνας εἰκόνων τὰ παρ' ἡμῖν διπλᾶ γράμματα ὑπάρχειν, ἃ συναριθμούμενα τοῖς εἰκοσιτέσσαρσι στοιχείοις δυνάμει τῇ κατὰ ἀναλογίαν τὸν τῶν τριάκοντα ποιεῖ ἀριθμόν.

(1.14.6) Τούτου τοῦ λόγου καὶ τῆς οἰκονομίας ταύτης καρπόν φησιν ἐν ὁμοιωματι εἰκόνος πεφηνέναι ἐκεῖνον τὸν μετὰ τὰς ἓξ ἡμέρας τέταρτον ἀναβάντα εἰς τὸ ὄρος καὶ γενόμενον ἕκτον, τὸν καταβάντα καὶ κρατηθέντα ἐν τῇ Ἑβδομάδι, ἐπίσημον Ὀγδοάδα ὑπάρχοντα καὶ ἔχοντα ἐν ἑαυτῷ τὸν ἅπαντα τῶν στοιχείων ἀριθμόν, ὃν ἐφανέρωσεν, ἐλθόντος αὐτοῦ ἐπὶ τὸ βάπτισμα, ἡ τῆς περιστερᾶς κάθοδος, ἥτις ἐστὶν ὦ καὶ ᾱ· ὁ γὰρ ἀριθμὸς αὐτῆς ἓν καὶ ὀκτακόσια. Καὶ διὰ τοῦτο Μωϋσέα ἐν τῇ ἕκτῃ ἡμέρᾳ εἰρηκέναι τὸν ἄνθρωπον γεγονέναι, καὶ τὴν οἰκονομίαν δὲ ἐν τῇ ἕκτῃ τῶν ἡμερῶν, ἥτις ἐστὶν ἡ παρασκευή, <ἐν> ᾗ τὸν ἔσχατον ἄνθρωπον εἰς ἀναγέννησιν τοῦ πρώτου ἀνθρώπου πεφηνέναι, ἧς οἰκονομίας ἀρχὴν καὶ τέλος τὴν ἕκτην ὥραν εἶναι, ἐν ᾗ προσηλώθη τῷ ξύλῳ. Τὸν γὰρ τέλειον Νοῦν, ἐπιστάμενον τὸν τῶν ἓξ ἀριθμὸν δύναμιν ποιήσεως καὶ ἀναγεννήσεως ἔχοντα, φανερῶσαι τοῖς υἱοῖς τοῦ φωτὸς τὴν διὰ τοῦ φανέντος ἐπισήμου εἰς αὐτὸν ἀριθμοῦ γενομένην ἀναγέννησιν. Ἔνθεν καὶ τὰ διπλᾶ γράμματα τὸν ἀριθμὸν ἐπίσημον ἔχειν φησίν· ὁ γὰρ ἐπίσημος ἀριθμὸς συγκεκρασθεὶς τοῖς εἰκοσιτέσσαρσι στοιχείοις τὸ τριακονταγράμματον ὄνομα ἀπετέλεσεν.

(1.14.7) Κέχρηται δὲ διακόνῳ τῷ τῶν ἑπτὰ ἀριθμῶν μεγέθει, ὥς φησιν ἡ Μάρκου Σίγη, ἵνα τῆς αὐτοβουλήτου βουλῆς φανερωθῇ ὁ καρπός. Τὸν μέντοι ἐπίσημον τοῦτον ἀριθμὸν ἐπὶ τοῦ παρόντος, φησί, τὸν ἐπὶ τοῦ ἐπισήμου μορφωθέντα νόησον, τὸν ὥσπερ μερισθέντα ἢ διχοτομηθέντα καὶ ἔξω μείναντα, ὃς τῇ ἑαυτοῦ δυνάμει τε καὶ φρονήσει διὰ τῆς ἀπ' αὐτοῦ προβολῆς τοῦτον τὸν τῶν ἑπτὰ δυνάμεων, κατὰ μίμησιν τῆς Ἑβδομάδος δυνάμεως ἐψύχωσε κόσμον καὶ ψυχὴν ἔθετο εἶναι τοῦ ὁρωμένου παντός. Κέχρηται μὲν οὖν καὶ αὐτὸς οὗτος τῷδε τῷ ἔργῳ ὡς αὐθαιρέτως ὑπ' αὐτοῦ γενομένῳ, τὰ δὲ διακονεῖ, μιμήματα ὄντα τῶν ἀμιμήτων, τὴν Ἐνθύμησιν τῆς Μητρός. Καὶ ὁ μὲν πρῶτος οὐρανὸς φθέγγεται τὸ α, ὁ δὲ μετὰ τοῦτον τὸ ε, ὁ δὲ τρίτος τὸ η, τέταρτος δὲ καὶ μέσος τῶν ἑπτὰ τὴν τοῦ ἰῶτα δύναμιν ἐκφωνεῖ, ὁ δὲ πέμπτος τὸ ο, ἕκτος δὲ τὸ υ, ἕβδομος δὲ καὶ τέταρτος ἀπὸ τοῦ μέσου τὸ ω στοιχεῖον ἐκβοᾷ, καθὼς ἡ Μάρκου Σιγή, ἡ πολλὰ μὲν φλυαροῦσα, μηδὲν δὲ ἀληθὲς λέγουσα, διαβεβαιοῦται. Αἵτινες Δυνάμεις ὁμοῦ, φησί, πᾶσαι εἰς ἀλλήλας συμπλακεῖσαι ἠχοῦσι καὶ δοξάζουσιν ἐκεῖνον ὑφ' οὗ προεβλήθησαν, ἡ δὲ δόξα τῆς ἠχῆς ἀναπέμπεται εἰς τὸν Προπάτορα. Ταύτης μέν τοι τῆς δοξολογίας τὸν ἦχον εἰς τὴν γῆν φερόμενόν φησι πλάστην γενέσθαι καὶ γεννήτορα τῶν ἐπὶ τῆς γῆς.

(1.14.8) Τὴν δὲ ἀπόδειξιν φέρει ἀπὸ τῶν ἄρτι γεννωμένων βρεφῶν, ὧν ἡ ψυχὴ ἅμα τῷ ἐκ μήτρας προελθεῖν ἐπιβοᾷ ἑνὸς ἑκάστου τῶν στοιχείων τούτων τὸν ἦχον. Καθὼς οὖν αἱ ἑπτά, φησί, Δυνάμεις δοξάζουσι τὸν Λόγον, οὕτως καὶ ἡ ψυχὴ ἐν τοῖς βρέφεσι κλαίουσα καὶ θρηνοῦσα Μάρκον δοξάζει αὐτόν. Διὰ τοῦτο δὲ καὶ τὸν Δαυὶδ εἰρηκέναι· "Ἐκ στόματος νηπίων καὶ θηλαζόντων κατηρτίσω αἶνον," καὶ πάλιν· "Οἱ οὐρανοὶ διηγοῦνται δόξαν Θεοῦ." Καὶ διὰ τοῦτο ἔν τε πόνοις καὶ ταλαιπωρίαις ψυχὴ γενομένη εἰς διϋλισμὸν αὐτῆς, ἐπιφωνεῖ τὸ ω εἰς σημεῖον αἰνέσεως, ἵνα γνωρίσασα ἡ ἄνω ψυχὴ τὸ συγγενὲς αὐτῆς βοηθὸν αὐτῇ καταπέμψῃ.

(1.15.1) Οὕτως οὖν ἀπαγγέλλει ἡ πάνσοφος αὐτῶν Σιγὴ τὴν γένεσιν τῶν εἰκοσιτεσσάρων στοιχείων· τῇ Μονότητι συνυπάρχειν Ἑνότητα, ἐξ ὧν δύο προβολαί, καθὰ προείρηται, Μονάς τε καὶ τὸ Ἕν, δὶς δύο οὖσαι τέσσαρες ἐγένοντο· δὶς γὰρ δύο τέσσαρες. Καὶ πάλιν αἱ δύο καὶ τέσσαρες εἰς τὸ αὐτὸ συντεθεῖσαι τὸν τῶν ἓξ ἐφανέρωσαν ἀριθμόν, οὗτοι δὲ οἱ ἓξ τετραπλασιασθέντες τὰς εἰκοσιτέσσαρας ἀπεκύησαν μορφάς. Καὶ τὰ μὲν τῆς πρώτης Τετράδος ὀνόματα, ἅγια ἁγίων νοούμενα καὶ μὴ δυνάμενα λεχθῆναι, γινώσκεσθαι ὑπὸ μόνου τοῦ Υἱοῦ, ἃ ὁ Πατὴρ οἶδεν τίνα ἐστι· τὰ δὲ σεμνῶς καὶ μετὰ πίστεως ὀνομαζόμενα παρ' αὐτῷ ἐστι ταῦτα· Ἄρρητος καὶ Σιγή, Πατήρ τε καὶ Ἀλήθεια. Ταύτης δὲ τῆς Τετράδος ὁ σύμπας ἀριθμός ἐστι στοιχείων εἰκοσιτεσσάρων. Τὸ γὰρ Ἄρρητος ὄνομα γράμματα ἔχει ἐν ἑαυτῷ ἑπτά, ἡ δὲ Σειγὴ πέντε, καὶ ὁ Πατὴρ πέντε, καὶ ἡ Ἀλήθεια ἑπτά· ἃ συντεθέντα ἐπὶ τὸ αὐτό, τὰ δὶς πέντε καὶ δὶς ἑπτά, τὸν τῶν εἰκοσιτεσσάρων ἀριθμὸν ἀνεπλήρωσεν. Ὡσαύτως δὲ καὶ ἡ δευτέρα Τετράς, Λόγος καὶ Ζωή, Ἄνθρωπος καὶ Ἐκκλησία, τὸν αὐτὸν ἀριθμὸν τῶν στοιχείων ἀνέδειξαν. Καὶ τὸ τοῦ Σωτῆρος δὲ ῥητὸν ὄνομα Ἰησοῦς γραμμάτων ὑπάρχει ἕξ, τὸ δὲ ἄρρητον αὐτοῦ γραμμάτων εἰκοσιτεσσάρων. Υἱὸς Χρειστὸς γραμμάτων δώδεκα, τὸ δὲ ἐν Χριστῷ ἄρρητον γραμμάτων τριάκοντα. Καὶ διὰ τοῦτό φησιν αὐτὸν α καὶ ω, ἵνα τὴν περιστερὰν μηνύσῃ, τοῦτον ἔχοντος τὸν ἀριθμὸν τούτου τοῦ ὀρνέου.

(1.15.2) Ὁ δὲ Ἰησοῦς τούτην ἔχει, φησί, τὴν ἄρρητον γένεσιν. Ἀπὸ γὰρ τῆς Μητρὸς τῶν ὅλων, τῆς πρώτης Τετράδος, ἐν θυγατρὸς τρόπῳ προῆλθεν ἡ δευτέρα Τετράς, καὶ ἐγένετο Ὀγδοάς, ἐξ ἧς προῆλθεν Δεκάς· οὕτως ἐγένετο Δεκὰς καὶ Ὀγδοάς. Ἡ οὖν Δεκὰς ἐπισυνελθοῦσα τῇ Ὀγδοάδι καὶ δεκαπλασίονα αὐτὴν ποιήσασα τὸν τῶν ὀγδοήκοντα προεβίβασεν ἀριθμόν, καὶ τὰ ὀγδοήκοντα πάλιν δεκαπλασιάσασα τὸν τῶν ὀκτακοσίων ἀριθμὸν ἐγέννησεν, ὥστε εἶναι τὸν ἅπαντα τῶν γραμμάτων ἀριθμὸν ἀπὸ Ὀγδοάδος εἰς Δεκάδα προελθόντα ῆ καὶ π̄ καὶ ῶ, ὅ ἐστιν Ἰησοῦς· τὸ γὰρ Ἰησοῦς ὄνομα κατὰ τὸν ἐν τοῖς γράμμασιν ἀριθμὸν ὀκτακόσιά ἐστιν ὀγδοήκοντα ὀκτώ. Ἔχεις σαφῶς καὶ τὴν ὑπερουράνιον τοῦ Ἰησοῦ κατ᾽ αὐτοὺς γένεσιν. Διὸ καὶ τὸν ἀλφάβητον τῶν Ἑλλήνων ἔχειν μονάδας ὀκτὼ καὶ δεκάδας ὀκτὼ καὶ ἑκατοντάδας ὀκτώ, τὴν τῶν ὀκτακοσίων ὀγδοήκοντα ὀκτὼ ψῆφον ἐπιδεικνύοντα, τουτέστι τὸν Ἰησοῦν, ἐκ πάντων συνεστῶτα τῶν ἀριθμῶν. Καὶ διὰ τοῦτο ἄλφα καὶ ω ὀνομάζεσθαι αὐτόν, τὴν ἐκ πάντων <αὐτοῦ> γένεσιν σημαίνοντα. Καὶ πάλιν οὕτως· τῆς πρώτης Τετράδος κατὰ πρόβασιν ἀριθμοῦ εἰς αὐτὴν συντιθεμένης, ὁ τῶν δέκα ἀνεφάνη ἀριθμός· μία γὰρ καὶ δύο καὶ τρεῖς καὶ τέσσαρες ἐπὶ τὸ αὐτὸ συντεθεῖσαι δέκα γίνονται, ὅ ἐστιν ῑ, καὶ τοῦτ᾽ εἶναι θέλουσι τὸν Ἰησοῦν. Ἀλλὰ καὶ ὁ Χρειστός, φησί, γραμμάτων ὀκτὼ ὤν, τὴν πρώτην Ὀγδοάδα σημαίνει, ἥτις τῷ ι συμπλακεῖσα τὸν Ἰησοῦν ἀπεκύησε. Λέγεται δέ, φησί, καὶ Υἱὸς Χρειστός, τουτέστιν ἡ Δωδεκάς· τὸ γὰρ Υἱὸς ὄνομα γραμμάτων ἐστὶ τεσσάρων, τὸ δὲ Χρειστὸς ὀκτώ, ἅτινα συντεθέντα τὸ τῆς Δωδεκάδος ἐπέδειξαν μέγεθος. Πρὶν μὲν οὖν, φησί, τούτου τοῦ ὀνόματος τὸ ἐπίσημον φανῆναι, τουτέστιν τὸν Ἰησοῦν, τοῖς υἱοῖς, ἐν ἀγνοίᾳ πολλῇ ὑπῆρχον οἱ ἄνθρωποι καὶ πλάνῃ· ὅτε δὲ ἐφανερώθη τὸ ἑξαγράμματον ὄνομα, ὃ σάρκα περιεβάλετο, ἵνα εἰς τὴν αἴσθησιν τοῦ ἀνθρώπου κατέλθῃ, ἔχον ἐν ἑαυτῷ αὐτὰ τὰ ἓξ καὶ τὰ εἰκοσιτέσσαρα, τότε γνόντες αὐτὸ ἐπαύσαντο τῆς ἀγνοίας, ἐκ θανάτου δὲ εἰς ζωὴν ἀνῆλθον, τοῦ ὀνόματος αὐτοῖς ὁδοῦ γενηθέντος πρὸς τὸν Πατέρα τῆς Ἀληθείας. Τεθεληκέναι γὰρ τὸν Πατέρα τῶν ὅλων λῦσαι τὴν ἄγνοιαν καὶ καθελεῖν τὸν θάνατον. Ἀγνοίας δὲ λύσις ἡ ἐπίγνωσις αὐτοῦ ἐγίνετο. Καὶ διὰ τοῦτο ἐκλεχθῆναι κατὰ τὸ θέλημα αὐτοῦ τὸν κατ᾽ εἰκόνα τῆς ἄνω Δυνάμεως οἰκονομηθέντα ἄνθρωπον.

(1.15.3) Ἀπὸ Τετράδος γὰρ προῆλθον οἱ Αἰῶνες. Ἦν δὲ ἐν τῇ Τετράδι Ἄνθρωπος καὶ [ἡ] Ἐκκλησία, Λόγος καὶ Ζωή. Ἀπὸ τούτων οὖν δυνάμεις, φησίν, ἀπορρυεῖσαι ἐγενεσιούργησαν τὸν ἐπὶ γῆς φανέντα Ἰησοῦν. Καὶ τοῦ μὲν Λόγου ἀναπεπληρωκέναι τὸν τόπον τὸν ἄγγελον Γαβριήλ, τῆς δὲ Ζωῆς τὸ ἅγιον Πνεῦμα, τοῦ δὲ Ἀνθρώπου τὴν τοῦ Ὑψίστου δύναμιν· τὸν δὲ τῆς Ἐκκλησίας τόπον ἡ Παρθένος ἐπέδειξεν. Οὕτως τε ὁ κατ᾽ οἰκονομίαν διὰ τῆς Μαρίας γενεσιουργεῖται παρ᾽ αὐτῷ ἄνθρωπος, ὃν ὁ Πατὴρ τῶν ὅλων διελθόντα διὰ μήτρας ἐξελέξατο διὰ Λόγου εἰς ἐπίγνωσιν αὐτοῦ. Ἐλθόντος δὲ αὐτοῦ εἰς τὸ ὕδωρ, κατελθεῖν εἰς αὐτὸν ὡς περιστερὰν τὸν ἀναδραμόντα ἄνω καὶ πληρώσαντα τὸν δωδέκατον ἀριθμόν, ἐν ᾧ ὑπάρχει τὸ σπέρμα τούτων τῶν συσπαρέντων αὐτῷ καὶ συγκαταβάντων καὶ συναναβάντων. Αὐτὴν δὲ τὴν δύναμιν τὴν κατελθοῦσαν σπέρμα φησὶν εἶναι τοῦ Πατρός, ἔχον ἐν ἑαυτῷ καὶ τὸν Πατέρα καὶ τὸν Υἱὸν τήν τε διὰ τούτων γινωσκομένην ἀνονόμαστον δύναμιν τῆς Σιγῆς καὶ τοὺς ἅπαντας Αἰῶνας. Καὶ τοῦτ' εἶναι τὸ Πνεῦμα τὸ λαλῆσαν διὰ τοῦ στόματος τοῦ Ἰησοῦ, τὸ ὁμολογῆσαν ἑαυτὸ Υἱὸν Ἀνθρώπου καὶ φανερῶσαν τὸν Πατέρα, κατελθὸν μὲν εἰς τὸν Ἰησοῦν, ἐνωθὲν δ' αὐτῷ. Καὶ καθεῖλε μὲν τὸν θάνατον, φησίν, ὁ ἐκ τῆς οἰκονομίας Σωτήρ, ἐγνώρισε δὲ τὸν Πατέρα Χριστόν. Εἶναι οὖν τὸν Ἰησοῦν ὄνομα μὲν τοῦ ἐκ τῆς οἰκονομίας ἀνθρώπου λέγει, τεθεῖσθαι δὲ εἰς ἐξομοίωσιν καὶ μόρφωσιν τοῦ μέλλοντος εἰς αὐτὸν κατέρχεσθαι Ἀνθρώπου, ὃν χωρήσαντα ἐσχηκέναι αὐτὸν αὐτόν τε τὸν Ἄνθρωπον αὐτόν τε τὸν Λόγον καὶ τὸν Πατέρα καὶ τὸν Ἄρρητον καὶ τὴν Σιγὴν καὶ τὴν Ἀλήθειαν καὶ Ἐκκλησίαν καὶ Ζωήν.

(1.16.1) Τὴν οὖν γένεσιν τῶν Αἰώνων αὐτῶν καὶ τὴν πλάνην τοῦ προβάτου καὶ ἀνεύρεσιν, ἐνώσαντες ἐπὶ τὸ αὐτό, μυστικώτερον ἐπιχειροῦσιν ἀπαγγέλλειν οὗτοι οἱ εἰς ἀπιθμοὺς τὰ πάντα κατάγοντες, ἐκ μονάδος καὶ δυάδος φάσκοντες τὰ ὅλα συνεστηκέναι. Καὶ ἀπὸ μονάδος ἕως τῶν τεσσάρων ἀριθμοῦντες οὕτω γεννῶσι τὴν Δεκάδα· μία γὰρ καὶ δύο καὶ τρεῖς καὶ τέσσαρες συντεθεῖσαι ἐπὶ τὸ αὐτὸ τὸν τῶν δέκα Αἰώνων ἀπεκύησαν ἀριθμόν. Πάλιν δ' αὖ ἡ δυὰς ἀπ' αὐτῆς προελθοῦσα ἕως τοῦ ἐπισήμου, οἷον δύο καὶ τέσσαρες καὶ ἕξ, τὴν Δωδεκάδα ἐπέδειξεν. Καὶ πάλιν ἀπὸ τῆς δυάδος ὁμοίως ἀριθμούντων ἡμῶν ἕως τῶν δέκα, ἡ Τριακοντὰς ἀνεδείχθη, ἐν ᾗ Ὀγδοὰς καὶ Δεκὰς καὶ Δωδεκάς. Τὴν οὖν Δωδεκάδα, διὰ τὸ [ἐπίσημον] ἐσχηκέναι συνεπακολουθῆσαν αὐτῇ τὸ ἐπίσημον, πάθος λέγουσι. Καὶ διὰ τοῦτο, περὶ τὸν δωδέκατον ἀριθμὸν τοῦ σφάλματος γενομένου, τὸ πρόβατον ἀποσκιρτῆσαν πεπλανῆσθαι, ἐπειδὴ τὴν ἀπόστασιν ἀπὸ Δωδεκάδος γεγενῆσθαι φάσκουσι. Τῷ αὐτῷ τρόπῳ καὶ ἀπὸ τῆς Δωδεκάδος ἀποστᾶσαν μίαν Δύναμιν ἀπολωλέναι μαντεύονται, καὶ ταύτην εἶναι τὴν γυναῖκα τὴν ἀπολέσασαν τὴν δραχμὴν καὶ ἅψασαν λύχνον καὶ εὑροῦσαν αὐτήν· Οὕτως οὖν καὶ τοὺς ἀριθμοὺς τοὺς καταλειφθέντας, ἐπὶ μὲν τῆς δραχμῆς τοὺς ἐννέα, ἐπὶ δὲ τοῦ προβάτου τοὺς ἕνδεκα, ἐπιπλεκομένους ἀλλήλοις τὸν τῶν ἐνενήκοντα ἐννέα τίκτειν ἀριθμόν, ἐπεὶ ἐννάκις τὰ ἕνδεκα ἐνενήκοντα ἐννέα γίνεται. Διὸ καὶ τὸ ἀμὴν τοῦτον λέγουσιν ἔχειν τὸν ἀριθμόν.

(1.16.2) Οὐκ ὀκνήσω δέ σοι καὶ ἄλλως ἐξηγουμένων αὐτῶν ἀπαγ-γεῖλαι, ἵνα πανταχόθεν κατανοήσῃς τὸν καρπὸν αὐτῶν. τὸ γὰρ στοιχεῖον τὸ η σὺν μὲν τῷ ἐπισήμῳ Ὀγδοάδα εἶναι θέλουσιν, ἀπὸ τοῦ ἄλφα ὀγδόῳ κείμενον τόπῳ· εἶτα πάλιν ἄνευ τοῦ ἐπισήμου ψηφίζοντες τὸν ἀριθμὸν αὐτῶν τῶν στοιχείων καὶ ἐπισυνθέντες μέχρι τοῦ ἦτα, τὴν Τριακοντάδα ἐπιδεικνύουσιν. Ἀρξάμενος γάρ τις ἀπὸ τοῦ ἄλφα καὶ τελευτῶν εἰς τὸ ἦτα τῷ ἀριθμῷ τῶν στοιχείων, ὑφεξαιρούμενος δὲ τὸ ἐπίσημον καὶ ἐπισυντιθεὶς τὴν ἐπαύξησιν τῶν γραμμάτων, εὑρήσει τὸν τῶν τριάκοντα ἀριθμόν. Μέχρι γὰρ τοῦ ε στοιχείου πεντεκαίδεκα γίνονται· ἔπειτα προστεθεὶς αὐτοῖς ὁ τῶν ἑπτὰ ἀριθμὸς β̄ καὶ κ̄ ἀπετέλεσε· προσελθὸν δὲ τούτοις τὸ η, ὅ ἐστιν ὀκτώ, τὴν θαυμασιωτάτην Τριακοντάδα ἀνεπλήρωσεν. Καὶ ἐντεῦθεν ἀποδεικνύουσι τὴν Ὀγδοάδα Μητέρα τῶν τριάκοντα Αἰώνων. Ἐπεὶ οὖν ἐκ τριῶν δυνάμεων ἥνωται ὁ τῶν τριάκοντα ἀριθμός, τρὶς αὐτὸς γενόμενος τὰ ἐνενήκοντα ἐποίησε· τρὶς γὰρ τριάκοντα ἐνενήκοντα. Καὶ αὐτὴ δὲ ἡ τριὰς ἐφ' ἑαυτὴν συντεθεῖσα ἐννέα ἐγέννησεν. Οὕτως τε ἡ Ὀγδοὰς τὸν τῶν ἐνενήκοντα ἐννέα παρ' αὐτοῖς ἀπεκύησεν ἀριθμόν. καὶ ἐπεὶ ὁ δωδέκατος Αἰὼν ἀποστὰς κατέλειψε τοὺς ἄνω ἕνδεκα, καταλλήλως λέγουσι τὸν τύπον τῶν γραμμάτων τῷ σχήματι τοῦ Λόγου κεῖται· ἐνδέκατον γὰρ τῶν γραμμάτων εἶναι τὸ λ, ὅ ἐστιν ἀριθμὸς τῶν τριάκοντα, καὶ κατ' εἰκόνα κεῖσθαι τῆς ἄνω οἰκονομίας, ἐπειδὴ ἀπὸ τοῦ ἄλφα, χωρὶς τοῦ ἐπισήμου, αὐτῶν τῶν γραμμάτων ὁ ἀριθμὸς ἕως τοῦ λ συντιθέμενος κατὰ τὴν παραύξησιν τῶν γραμμάτων σὺν αὐτῷ τῷ λ τὸν τῶν ἐνενήκοντα ἐννέα ποιεῖται ἀριθμόν. Ὅτι δὲ τὸ λ ἐνδέκατον ὂν τῇ τάξει ἐπὶ τὴν τοῦ ὁμοίου αὐτῷ κατῆλθεν ζήτησιν, ἵνα ἀναπληρώσῃ τὸν δωδέκατον ἀριθμόν, καὶ εὑρὸν αὐτὸ ἐπληρώθη, φανερὸν εἶναι ἐξ αὐτοῦ τοῦ σχήματος τοῦ στοιχείου. Τὸ γὰρ λ, ὥσπερ ἐπὶ τὴν τοῦ ὁμοίου αὐτῷ ζήτησιν παραγενόμενον καὶ εὑρὸν καὶ εἰς ἑαυτὸ ἁρπάσαν αὐτό, τὴν τοῦ δωδεκάτου ἀνεπλήρωσεν χώραν, τοῦ Μ στοιχείου ἐκ δύο Λ συγκειμένου. Διὸ καὶ φεύγειν αὐτοὺς διὰ τῆς γνώσεως τὴν τῶν ἐνενήκοντα ἐννέα χώραν, τουτέστιν τὸ ὑστέρημα, τύπον ἀριστερᾶς χειρός, μεταδιώκειν δὲ τὸ ἕν, ὃ προστεθὲν τοῖς ἐνενήκοντα ἐννέα εἰς τὴν δεξιὰν αὐτοὺς χεῖρα μετέστησεν.

Proclus *Commentary on the "Timaeus,"* Testimony 6

Chapter 8 above (pp. 159–161):

Θεόδωρος δὲ ὁ ἐκ τῆς Ἀσίνης φιλόσοφος, τῶν Νουμηνείων λόγων ἐμφορηθείς, καινοπρεπέστερον τοὺς περὶ τῆς ψυχογονίας διέθηκε λόγους, ἀπὸ τῶν γραμμάτων καὶ τῶν χαρακτήρων καὶ τῶν ἀριθμῶν ποιούμενος τὰς ἐπιβολάς. ἵνα οὖν καὶ τὰ τούτῳ δοκοῦντα συντόμως ἔχωμεν ἀναγεγραμμένα, φέρε, καθ' ἕκαστον ὧν λέγει, ποιησώμεθα σύνοψιν ἐν κεφαλαίοις.

τὸ μὲν οὖν πρῶτον ἄρρητον αὐτῷ καὶ ἀνεκλάλητον καὶ πηγὴ τῶν πάντων καὶ τῆς ἀγαθότητος αἴτιον καλῶς ἀνύμνηται. μετὰ δὲ τοῦτο ⟨τὸ⟩ οὕτως ἐξηρημένον τῶν ὅλων τριάς ἐστιν ἢ τὸ νοητὸν αὐτῷ πλάτος ὁρίζουσα, ἣν καλεῖ τὸ ἕν, ἔκ τε τοῦ ἄσθματος οὖσαν τοῦ ἀρρήτου πως ὄντος, ὃ μιμεῖται ἡ δασεῖα τοῦ ἕν, καὶ ἐκ τῆς ἀψῖδος αὐτοῦ τοῦ ε̄ μόνου, χωρὶς τοῦ συμφώνου, καὶ αὐτοῦ ἤδη τοῦ ν̄. ἄλλη δὲ μετὰ ταύτην τριὰς ὁρίζει τὸ νοερὸν βάθος καὶ ἄλλη τὸ δημιουργικόν· ἡ μὲν γάρ ἐστι τὸ εἶναι πρὸ τοῦ ὄντος, τὸ νοεῖν πρὸ τοῦ νοῦ, τὸ ζῆν πρὸ τῆς ζωῆς. ἡ δὲ δημιουργικὴ τριὰς μετὰ ταύτας ἐστί, πρῶτον μὲν ἔχουσα τὸ ὄν, δεύτερον δὲ τὸν νοῦν, τρίτον δὲ τὴν πηγὴν τῶν ψυχῶν,

ἀπὸ δὲ ταύτης τῆς τριάδος ἄλλη τριάς, ἡ αὐτοψυχὴ καὶ ἡ καθόλου καὶ ἡ τοῦ παντός, περὶ ὧν τῆς διαιρέσεως καὶ ἔμπροσθεν εἴπομεν, ὧν ἑκάστη προῆλθε μὲν ἀπὸ πάσης τῆς δημιουργικῆς τριάδος, ἀλλὰ μᾶλλον ἢ μὲν ἀπὸ τοῦ ὄντος, ἢ δὲ ἀπὸ τοῦ νοῦ, ἢ δὲ ἀπὸ τῆς πηγαίας ψυχῆς. περὶ δὴ ταύτης τῆς τοῦ παντὸς ψυχῆς τῷ Πλάτωνι προκεῖσθαι λέγειν, μᾶλλον δὲ περὶ τῆς ἁπλῶς ψυχῆς τῆς ἀπὸ τῆς πηγῆς τῶν ψυχῶν καὶ περὶ τῆς καθόλου καὶ τῆς τοῦ παντὸς καὶ περὶ αὐτῆς τῆς πηγῆς· εἶναι γὰρ πάντα ἐν πᾶσιν, εἰ καὶ ὅπου μὲν ἄλλως, ὅπου δὲ ἄλλως,

καὶ ἐν μὲν τῇ πρὸ τῆς τριάδος ψυχῇ καθ' ἕνωσιν,

ἐν δὲ τῇ ἁπλῶς καθ' ὁλότητα τὴν πρὸ τῶν μερῶν,

ἐν δὲ τῇ καθόλου κατὰ τὴν ἐκ τῶν μερῶν ὁλότητα,

καὶ ἐν τῇ τρίτῃ κατὰ τὴν ἐν τοῖς μέρεσιν,

ὡς τοῦ Πλάτωνος περὶ πασῶν διατατττομένου τούτων καὶ δέοντος εἰς πάσας ἀνάγειν τοὺς λόγους ἅπαντας, μηδὲ τὴν διαφορὰν αὐτῶν παριέντας.

καὶ πρῶτον δι' ἣν αἰτίαν ἐκ τριῶν ἐστι μεσοτήτων, οἴεται δεῖν λέγειν. καὶ δὴ καί φησιν, ὅτι τὸ μὲν ὅλον γεωμετρικός ἐστιν ἡ ψυχὴ λόγος, ὑποστὰς ἔκ τε τοῦ πρώτου θεοῦ τοῦ κατὰ τὸ ὂν καὶ τοῦ δευτέρου τοῦ κατὰ τὸν νοῦν· αὗται γὰρ αἱ δύο οὐσίαι, ἡ ἀμέριστος καὶ ἡ μεριστή· τελοῦσι δὲ εἰς αὐτὸν ὅ τε ἀριθμητικὸς λόγος εἰκόνα φέρων τῆς πρώτης οὐσίας καὶ ὁ ἁρμονικὸς τῆς δευτέρας· ὃ μὲν γάρ ἐστι μοναδικός, ἀδιάστατος ὤν, ὃ δὲ διεστὼς μέν, ἀλλὰ ἁρμονικῶς.

ἔπειθ᾽, ὅτι ἀπὸ μὲν τῆς τετράδος τῶν στοιχείων δείκνυται οὖσα τετράς, καὶ ὁ σύμπας ἀριθμὸς γεωμετρικὸς ἂν εἴη τις ἀριθμός. ἵνα δὲ μὴ ἄζων τοῦτον ὑπολάβῃς τὸν ἀριθμόν, ἐν τοῖς ἄκροις γράμμασιν εὑρήσεις τὸ ζῆν, ἀντὶ τῆς τρίτης ἑπτάδος τὴν πρώτην λαβών.

μᾶλλον δὲ ἐκθέμενος τοὺς πυθμένας τοῦ πρώτου γράμματος ἐφεξῆς αὐτῷ ζωὴν νοερὰν οὖσαν θεάσῃ τὴν ψυχήν· οἷον ζ̄ ō ψ̄· μέσος δὲ ὁ κύκλος, ὁ νοερὸς ὤν, ὅτι νοῦς αἴτιος τῆς ψυχῆς. καὶ ὁ μὲν ἐλάχιστος γεωμετρικὸν αὐτὴν ἀποφαίνει τινὰ νοῦν διὰ τῆς τῶν παραλλήλων συναφῆς καὶ τῆς διαμετρικῆς εὐθείας, ἄνω τε μένοντα καὶ ἐπὶ τἀναντία χωροῦντα καὶ ἀπλαγίαστον ἅμα καὶ πεπλαγιασμένον ζωῆς εἶδος ἐπιδεικνύμενον. ὁ δὲ μέγιστος στοιχεῖόν ἐστι σφαίρας· κατακαμφθεῖσαι γοῦν αἱ γραμμαὶ τὴν σφαῖραν ποιήσουσιν. ἐπὶ δὲ τούτῳ τοῦ ἑξῆς γράμματος οἱ πυθμένες, δ̄ μ̄ ū, τρεῖς τε ὄντες πάλιν καὶ τετραδικοὶ καὶ διὰ τοῦτο τὸν δώδεκα γεννῶντες, τὰς δώδεκα σφαίρας ἀποτελοῦσι τοῦ παντός. αὐτὸς δὲ ὁ μέγιστος τῶν πυθμένων δείκνυσι τὴν οὐσίαν αὐτῆς δυοῖν τινων ὀρεγομένην καὶ πρὸς δύο πράγματα ἀνατεινομένην· διὸ καὶ τοῦτο τὸ γράμμα τινὲς καλοῦσι φιλόσοφον· *** αὐτὴν δὲ εἰς τὸ κάτω ῥυεῖσαν ἀμφοτέρων. οὕτω γοῦν εὕρομεν αὐτὸ καὶ ἡμεῖς τὸ ū παρά τισι τῶν περιττῶν ὀνομαζόμενον, μεταξύ τέ ἐστι δύο σφαιρῶν, τοῦ τε ψ̄ καὶ τοῦ χ̄, τούτου μὲν θερμοτέρου διὰ τὸ πνεῦμα καὶ ζωτικωτέρου μᾶλλον ὄντος, ἐκείνου δὲ ἧττον ἑκάτερον ἔχοντος· ὥστε πάλιν δύο νόων ἐστὶ μεσότης, τοῦ μὲν προτέρου, τοῦ δὲ ὑστέρου, καὶ ὁ μέσος χαρακτὴρ τὴν πρὸς ἑκάτερον αὐτῆς οἰκειότητα καὶ σχέσιν δηλοῖ. μᾶλλον μὴν ὁ Πλάτων τὸ χ̄ τῇ ψυχῇ δέδωκε, καίτοι καὶ τοῦ ψ̄ στοιχείου σφαίρας ὄντος, ἵνα τὸ ἰσορρεπὲς ἐνδείξηται τῆς κινήσεως αὐτῆς, πασῶν τῶν εὐθειῶν ἴσων οὐσῶν ἐν τῷ χ̄, καὶ οὕτω τὸ αὐτοκίνητον γνώριμον ποιήσῃ τῆς ψυχῆς. εἰ δὲ αὐτῷ τῷ εἶναι παράγει τὴν ψυχὴν ὁ δημιουργός, δῆλον, ὅτι καὶ αὐτὸς ἐν ἀναλογίᾳ τέτακται τοῦ χ̄· τοῦτο γάρ ἐστιν ὁ πρώτιστος νοῦς. διὰ μὲν οὖν τούτων φησίν, ὅτι μέση τίς ἐστιν οὐσία δύο νόων ἡ ψυχὴ προϊοῦσα καὶ ἑαυτὴν παράγουσα. καὶ ταῦτα μὲν οὕτω ληπτέον. διὰ δὲ τοῦ τελευταίου γράμματος, τοῦ η̄, θεατέον τὴν ἄχρι τοῦ κύβου πρόοδον αὐτῆς.

εἰ δὲ διὰ τὴν ἑτερότητα τῆς ζωῆς δυάς ἐστι, διὰ δὲ τὸ τριμερὲς τῆς οὐσίας αὐτῆς τριάς, αὐτόθεν μὲν ἔχει τὸν ἡμιόλιον λόγον, εἰς ἑαυτὴν δὲ εἰσιοῦσα καὶ διὰ τὴν εἴσοδον ταύτην ἐπὶ τὴν τριάδα ποιοῦσα τὴν δυάδα γεννᾷ μὲν τὴν ἑξάδα, συνάπτουσα δὲ τῷ τε ἀμερίστῳ καὶ τῷ τριμερεῖ τὴν ἐν διπλασίῳ διὰ τούτων ἁρμονίαν ὑφίστησι. καὶ ὡς μὲν τριὰς εἰς ἑαυτὴν ἐπιστρέφουσα τὴν ἐννεάδα, ὡς δὲ δυὰς δυαδικῶς εἰς ἑαυτὴν ἰοῦσα τὴν ὀκτάδα πάντως, ἐκ δὲ ἀμφοῖν τὸν ἐπόγδοον λόγον ἀποτελεῖ.

καὶ ἡ μὲν κατὰ γραμμὴν ἀπογέννησις τό τε ἀμέριστον αὐτῆς δηλοῖ καὶ τὸ δι' ὅλου ταυτόν· πᾶν γὰρ μόριον γραμμῆς γραμμή· καὶ τὸ πάντας εἶναι πανταχοῦ τοὺς λόγους· ἡ δὲ κατὰ δύο σχίσις ὅτι δυαδικόν ἐστι τὸ εἶδος αὐτῆς ἐπιδείκνυσι. καὶ ἡ μὲν ὁλότης αὐτῆς ἡ ἀδιαίρετος εἰκών ἐστι τοῦ πρώτου νοῦ, ἡ δὲ ἄσχιστος τῶν δύο, ἣν ἐκάλεσε ταὐτοῦ κύκλον, τοῦ δευτέρου· ἡ δὲ εἰς ἓξ σχιζομένη τοῦ τρίτου τοῦ λοιποῦ διηριθμημένου. καὶ ἡ μὲν ὀκτὰς ὡς ἀπὸ δυάδος ἀνεφάνη τῆς ψυχῆς· ἡ δὲ ἑπτάς, ἣ μὲν ἐν μονάσι τὸ πρῶτον ἐνεικονίζεται τῆς ζωῆς εἶδος, ἣ δὲ ἐν δεκάσι τὸ νοερόν, διὰ τὸν κύκλον, ἣ δὲ ἐν ἑκατοντάσι τρίτον κατ' αὐτὸ λοιπὸν τὸ ἰδίωμα τὸ ψυχικόν. καὶ ἡ μὲν ἀπαρέγκλιτος αὐτῆς σύμφυσις πρὸς τὸ γεννῆσαν ὑπέστησε τὸ ἀπλανές, ἡ δὲ ἔξοδος καὶ ἡ ἀοριστία τὸ πλανώμενον, ἡ δὲ ἐπιστροφὴ μετὰ τὴν πρόοδον τὴν ἀπλανῶς πλανωμένην ζωήν. καὶ ἐπειδὴ τὸ μὲν σχῆμα τῆς ψυχῆς ἐστιν οἷον χ, τὸ δὲ εἶδος δυαδικόν (εἰς δύο γὰρ ἡ σχίσις), ἡ δὲ δυὰς ἐπὶ τὴν ἑξάδα πρῶτον οὖσαν πυθμένα τοῦ χ τὴν δυωδεκάδα ποιεῖ, λάβοις ἂν καὶ ἀπὸ τούτου τὰς δώδεκα πρώτας ἀρχικὰς ψυχάς.

Chapter 8 above (pp. 165–166):

ὁ μὲν οὖν Θεόδωρος τοιαῦτα ἄττα φιλοσοφεῖ περὶ τούτων, ἀπὸ τῶν γραμμάτων καὶ τῶν ἐκφωνήσεων τὰς ἐξηγήσεις ποιούμενος, ὡς ἐκ πολλῶν ὀλίγα παραθέσθαι. ὁ δέ γε θεῖος Ἰάμβλιχος ἅπασαν τὴν τοιαύτην θεωρίαν ἐπερράπισεν ἐν ταῖς πρὸς τοὺς ἀμφὶ Ἀμέλιον—οὕτω γὰρ ἐπιγράφει τὸ κεφάλαιον—καὶ δὴ καὶ Νουμήνιον ἀντιρρήσεσιν, εἴτε τοῦτον εἰς ἐκείνους ἀναπέμπων εἴτε καὶ ἐκείνοις ἐντυχών που τὰ ὅμοια γράφουσι περὶ τούτων· οὐ γὰρ ἔχω λέγειν.

λέγει δ' οὖν ὁ θεῖος Ἰάμβλιχος πρῶτον μέν, ὡς οὐκ ἔδει διὰ τὸ πλῆθος τῶν γραμμάτων τὴν ψυχὴν ποιεῖν τὸν σύμπαντα ἀριθμὸν ἢ τὸν γεωμετρικὸν ἀριθμόν· καὶ γὰρ τὸ σῶμα ἐκ τῶν ἴσων ἐστὶ γραμμάτων, καὶ αὐτὸ τὸ μὴ ὄν· ἔσται οὖν καὶ τὸ μὴ ὂν ἀριθμὸς ὁ σύμπας· πολλὰ δ' ἂν καὶ ἄλλα εὕροις ἐκ τῶν ἴσων ὄντα γραμμάτων καὶ αἰσχρὰ καὶ τὰ ἐναντιώτατα ἀλλήλοις, ἃ δὴ πάντα συγχεῖν εἰς ἄλληλα καὶ φύρειν οὐκ ὀρθῶς ἔχει.

δεύτερον δέ, ὅτι τὸ ἀπὸ τῶν χαρακτήρων ἐπιχειρεῖν οὐκ ἀσφαλές· θέσει γάρ ἐστι ταῦτα, καὶ τὸ μὲν ἀρχαῖον ἄλλος ἦν ὁ τύπος, νῦν δὲ ἄλλος· αὐτίκα τὸ ζ, ἐφ' οὗ πεποίηται τὸν λόγον ἐκεῖνος, οὔτε παραλλήλους εἶχε τὰς ἀπεναντίον πάντως οὔτε τὴν μέσην λοξήν, ἀλλὰ πρὸς ὀρθάς, ὡς καὶ ἀπὸ τῶν στηλῶν ἐστι τῶν ἀρχαίων καταφανές.

τρίτον τὸ εἰς τοὺς πυθμένας ἀναλύειν καὶ περὶ ἐκείνους διατρίβειν ἀπ' ἄλλων ἀριθμῶν εἰς ἄλλους μεθίστησι τὴν θεωρίαν· οὐ γὰρ ταὐτόν ἐστιν ἡ ἐν μονάσιν ἑπτὰς καὶ ἡ ἐν δεκάσι καὶ ἡ ἐν ἑκατοντάσι. ταύτης οὖν ἐν τῷ ὀνόματι τῆς ψυχῆς οὔσης τί ἔδει παρεισκυκλεῖν τὸν περὶ τῶν πυθμένων λόγον; οὕτω γὰρ ἂν πάντα εἰς πάντας ἀριθμοὺς μεταβάλλοιμεν διαιροῦντες ἢ συντιθέντες ἢ πολλαπλασιάζοντες.

καθόλου μὲν οὖν ταῦτα. διελέγχει δὲ καὶ ἑκάστην ἀπόδοσιν ὡς ἐσκευωρημένην καὶ οὐδὲν ἔχουσαν ὑγιές. καὶ ὅτῳ φίλον πάντων τὴν σαθρότητα κατανοῆσαι, ῥᾴδιον παραθεμένῳ τὸ βιβλίον ἀναλέγεσθαι τὰς οἰκείας ἀντιλογίας πρὸς ἕκαστον ἀπὸ τῶν γεγραμμένων.

Bibliography

Editions and Translations

Aelius Herodianus: *see* pseudo-Herodian.

Aetius *Placita*: Diels 1879.

Alexander of Aphrodisias *Commentary on Aristotle's "Metaphysics"*: Hayduck 1891.

Alexander Polyhistor, fragments and testimonies (including *Successions of the Philosophers*): Müller 1849:210–244.

Anatolius of Laodicea *On the Computation of the Pasch*: McCarthy and Breen 2003.

———. *On the Decad*: Heiberg 1901.

Anthologia Graeca: Beckby 1965.

Apocryphon of John (NH 4.1) (*see also* Nag Hammadi): Waldstein and Wisse 1995.

Apollonius Dyscolus: *see* Dionysius Thrax.

Apophasis Megale: *see* Hippolytus *Refutation of All Heresies*.

Archytas of Tarentum, fragments and testimonies: Diels and Kranz 1957; Huffman 2005.

Aristotle, fragments: Rose 1886.

———. *Metaphysics*: Ross 1924.

———. *On the Generation of Animals*: Dressaart Lulofs 1965.

———. *Problemata*: Bekker 1870:859–967.

Artemidorus *Dream Book*: Pack 1963.

Asclepius of Tralles *Commentary on Nicomachus of Gerasa, "Introduction to Arithmetic"*: Tarán 1969.

Athenagoras *Legatio*: Schoedel 1972:2–86.

pseudo-Barnabas *Epistle*: Kraft 1971.

Boethius *De unitate et de uno*: *Patrologia Latina* 63:1075–1078.

Books of Jeu: Schmidt 1978a.
Bruce codex (untitled treatise): see *Books of Jeu*.

Chalcidius *Commentary on the "Timaeus"*: Moreschini 2003.
Chrysippus, fragments: *Stoicorum Veterum Fragmenta* 2.
Clement of Alexandria *Epitomes*: Casey 1934.
————. *The Instructor*: Marrou, et al. 1960–1970.
————. *Stromateis*: Descourtieux 1999 (Book 8); Hort and Mayor 1902 (Book 7); Mondésert 1954 (Book 2); Stählin, et al. 1972, 1985, 1979 (complete); van den Hoek 2001 (Book 4); Ante-Nicene Fathers 2:299–567 (Eng. trans.).
Clement of Rome *Letter to the Corinthians*: Jaubert 1971.
Cyril of Alexandria *Against Julian*: Burguière and Évieux 1985.

Damascius *Commentary on "Parmenides"*: Ruelle 1899:2.5–322.
Democritus: *see* Presocratics.
Diogenes Laertius *Lives of the Philosophers*: Long 1964.
Dionysius Thrax (with scholia by Apollonius Dyscolus, Melampous, and other anonymous authors): Uhlig and Hilgard 1883–1901.

Epiphanius *Panarion*: F. Williams 1987 (English trans.); Holl 1922.
Eugnostos (NH 3.3): Parrott 1991.
Eunapius *Lives of the Sophists*: Giangrande 1956.
Eusebius *Church History*: Bardy 1952–1958.
Evagrius of Pontus *On Prayer*: Patrologia Graeca 79:1165–1200.

Favonius Eulogius *Disputation on the Dream of Scipio*: van Weddingen 1957.
First Apocalypse of James (NH 5.3): Kasser and Wurst 2007:115–161; Schoedel and Parrott 1979:65–103.

Galen *De diebus decretoriis*: Kühn 1825:769–941.
Gospel of Judas: Kasser and Wurst 2007:177–235.
Gospel of Philip (NH 2.3): Schenke 1997.
The Gospel of Truth (NH 1.3): Attridge and MacRae 1985.
Grammatici Graeci: *see* Dionysius Thrax.
Gregory Palamas *Sermons*: Chrestou 1985.

Heracleon, fragments: Wucherpfennig 2002.
Hermas *The Shepherd of Hermas*: Whittaker 1967.
Hermetic fragments: Nock 1983.
pseudo-Herodian Περὶ ἀριθμῶν: Estienne 1851:345.
Hierocles *On the "Golden Poem"*: Köhler 1974.
Hippocrates *On Hebdomads*: Roscher 1913.
Hippolytus *Refutation of All Heresies*: Marcovich 1986; ANF 5:9–153.

Iamblichus (see also *Theology of Arithmetic, On the Soul*):   Dillon 1973.

————. *Common Mathematical Knowledge*:   Festa 1891.

————. *On the Soul*:   Finamore and Dillon 2002.

————. *The Pythagorean Way of Life*:   Dillon and Hershbell 1991.

Interpretation of Knowledge (NH 11.1):   Hedrick 1990.

Irenaeus of Lyon, *Against Heresies*:   Rousseau et al. 1965–1982 (ed.); Unger and Dillon 1992 (Eng. trans.).

————. Fragments: Harvey 1857:2.470–511; Jordan 1913; Ter Měkěrttschian and Wilson 1919:732–744.

————. *Proof of the Apostolic Preaching*:   Ter Měkěrttschian and Wilson 1919; Behr 1997.

Jerome *Letters*:   Hilberg 1996.

John Lydus *On the Months*:   Wünsch 1898.

Julian *To the Untaught Dogs*:   Rochefort 1963:2.1:144–173.

Justin Martyr *Dialogue with Trypho*:   Goodspeed 1915.

pseudo-Justin Martyr *Exhortation to the Nations*:   Otto 1879.

Lamprias (*see also* Plutarch *Catalogue*):   Sandbach 1969:3–29.

Lucian *De lapsu inter salutandum*:   Kilburn 1959:172–188.

————. *Vitarum auctio*:   Harmon 1915:450–510.

"Lysis":   *see* Pythagorean texts.

Magical texts: Priesendanz and Heinrichs 1973–1974.

Marsanes (NH 10.1):   Funk et al. 2000; Pearson 1981:211–347.

Melampous: *see* Dionysius Thrax.

Moderatus of Gades, fragments: Mullach 1867:48–49.

The Mystery of the Letters: Bandt 2007.

Nag Hammadi (*see also individual treatises*):   Layton 1987; Robinson 1988.

Nemesius *On the Nature of Man*:   Morani 1987.

New Testament: Nestle, et al. 2004.

Nicomachus of Gerasa *Introduction to Arithmetic*:   Hoche 1866; D'Ooge 1926.

Numenius, fragments: des Places 1974.

On Simon: *see* Irenaeus *Against Heresies*.

[*On the Numbers*]:   Delatte 1915:171–175.

Origen *Against Celsus*:   Borret 1967–1976.

————. *Commentary on John*:   Blanc 1966–1992; Heine 1989–1993 (Eng. trans.).

Philo *On Abraham* (*De Abr.*); *Allegorical Interpretation* (*Leg. all.*); *On the Contemplative Life* (*De vita cont.*); *On the Creation of the World* (*De opif. mundi*); *On the Decalogue* (*De dec.*); *On Flight and Finding* (*De fuga et inv.*);

Philo (*cont.*) *On the Preliminary Studies* (*De congr. erud. gratia*); *On Rewards and Punishments* (*De praem. et poen.*); *On the Unchangeableness of God* (*Quod Deus sit immut.*); *Who Is the Heir of Divine Things?* (*Quis rerum div. heres*): Cohn and Wendland 1896–1930.

———. [*On Numbers*]: Stähle 1931.

———. *Questions and Answers on Genesis* (*Quaest. in Gen.*): Petit 1978; Mercier 1979.

Philolaus, fragments and testimonies: Huffman 1993.

Photius *Biblioteca*: Bekker 1825; Henry 1959–1991.

Pistis Sophia: Schmidt 1978b.

Plato *Parmenides; Philebus; Republic; Timaeus*: Burnet 1900–1907.

Pliny the Elder *Natural History*: Rackham 1938–1963.

Plotinus *Enneads*: Henry and Schwyzer 1951–1982; Armstrong 1984.

Plutarch *Genesis of the Soul in the "Timaeus"*: Hubert 1954.

———. *Isis and Osiris*: Gwyn Griffiths 1970.

———. *The Obsolescence of Oracles; On the E at Delphi*: Sieveking 1929.

———. *On Superstition*: Babbitt 1928.

———. *Table Talk*: Hubert 1938.

Porphyry *Commentary on Ptolemy's "Harmonics"*: Düring 1932:3–174.

———. *Life of Plotinus*: Henry and Schwyzer 1951:1:1–41.

———. *Life of Pythagoras*: Nauck 1886:17–52.

———. *On the Soul*: Bidez 1913.

Presocratics (Archytas, Democritus, Philolaus, Solon): Diels and Kranz 1957.

Proclus *Commentary on the First Book of Euclid's "Elements"*: Friedlein 1873.

———. *Commentary on the "Timaeus"*: Festugière 1966–1968; Taylor 1820.

Ptolemy [Valentinian] *Letter to Flora*: Quispel 1966.

[Claudius] Ptolemy *Harmonics*: Düring 1930.

pseudo-Pythagoras *Golden Poem*: Diehl 1971:86–94.

Pythagorean texts (pseudo-Archytas, "Lysis"): Diels and Kranz 1957; Thesleff 1965.

Revelation to Marcus: *see* Irenaeus *Against Heresies*.

Scholia in Dionysius Thrax: *see* Dionysius Thrax.

Scholia on Homer: Heyne 1834.

Septuagint: Rahlfs 1935.

Sextus Empiricus *Against the Logicians* (= *Against the Mathematicians* 6–7); *Against the Physicians; Outlines of Pyrrhonism*: Mau and Mutschmann 1914.

Simon Magus: *see* Hippolytus *Refutation of All Heresies*.

Simplicius *Commentary on Aristotle's "Physics"*: Diels 1882–1895.

Solon: *see* Presocratics.

Speusippus of Athens (*see also* Presocratics, fragments and testimonies): Tarán 1981.

Stobaeus *Eclogae*: Wachsmuth and Hense 1884–1912.

Syrianus *Commentary on Aristotle's "Metaphysics"*: Kroll 1902.

Tacitus *Annales*: Heubner 1983.

Tertullian *Against the Valentinians*: Fredouille 1980–1981; Riley 1971.

———. *On the Soul*: Waszink 1947.

The Tetraktys Suspending and Apportioning All Things Four-fold: Delatte 1915:187.

Theodore of Asine, testimonies: Deuse 1973.

Theodoret *Compendium of Heretical Fables*: Patrologia Graeca 83:336–556.

———. *Letters*: Azéma 1955–1998.

Theology of Arithmetic: de Falco 1922 (ed.); Waterfield 1989 (Eng. trans.).

Theon of Smyrna *Mathematics Useful for Reading Plato*: Hiller 1878.

Theophilus of Antioch *To Autolycus*: Grant 1970.

Treatise on the Resurrection (NH 1.4): Peel 1985.

The Tripartite Tractate (NH 1.5): Attridge and Pagels 1985; Thomassen 1989.

A Valentinian Exposition (NH 11.2): Robinson 1973; Turner 1990:89–172.

pseudo-Valentinus (Irenaeus *Against Heresies* 1.11.1): *see* Irenaeus.

Xenocrates, fragments and testimonies: Isnardi Parente 1982.

Zeno, fragments: *Stoicorum Veterum Fragmenta* 1.

Works Cited

Annas, J. 1976. *Aristotle's Metaphysics: Books M and N.* Oxford.

Apatow, R. 1999. "The Tetraktys: The Cosmic Paradigm of the Ancient Pythagoreans." *Parabola* 24.3:38–43.

Armstrong, A. H. 1992. "Dualism." In Wallis 1992:33–54.

———, ed. and trans. 1984. *Plotinus.* Vols. 4–5. LCL 443–444. Cambridge, MA.

Attridge, H., and G. W. MacRae, eds. 1985. "The Gospel of Truth." *Nag Hammadi Codex I (the Jung Codex)* 22:55–122; 23:38–135. NHS 22–23. Leiden.

Attridge, H., and E. Pagels, eds. 1985. "The Tripartite Tractate." *Nag Hammadi Codex I (the Jung Codex)* 22:159–337, 23:217–497. NHS 22–23. Leiden.

Azéma, Y., ed. 1955–1998. *Correspondance.* 4 vols. SC 40, 98, 111, 429. Paris.

Babbitt, F. C., ed. 1928. *Moralia.* II. Loeb Classical Library 222. Cambridge, MA.

Bandt, C., ed. and trans. 2007. *Der Traktat "Vom Mysterium der Buchstaben": Kritischer Text mit Einführung, Übersetzung und Anmerkungen.* Texte und Untersuchungen zur Geschichte der altchristlichen Literatur 162. Berlin.

Bardy, G., ed. 1952–1958. *Histoire ecclésiastique.* 3 vols. SC 31, 41, 55. Paris.

Beckby, H., ed. 1965. *Anthologia Graeca.* 4 vols. 2nd ed. Munich.

Behr, J. 2000. *Asceticism and Anthropology in Irenaeus and Clement.* Oxford.

———. 1997. *On the Apostolic Preaching.* Crestwood, NY.

Bekker, I. 1825. *Photii bibliotheca.* Vol. 2. Berlin.

———, ed. 1870. *Aristotelis opera.* Vol. 5. Berlin.

Bidez, J., ed. 1913. *Vie de Porphyre, le philosophe néo-platonicien, avec les fragments des traités Perì agalmátōn et De regressu animae.* Leipzig.

Bitton-Ashkelony, B. 2007. "Counseling through Enigmas: Monastic Leadership and Linguistic Techniques in Sixth-Century Gaza." *The Poetics of Grammar and the Metaphysics of Sound and Sign* (ed. S. La Porta and D. Shulman) 177–199. Leiden.

Blanc, C., ed. 1966–1992. *[Origen,] Commentaire sur saint Jean.* 5 vols. SC 120, 157, 222, 290, 385. Paris.

Bonazzi, M. 2007. "Eudorus of Alexandria and Early Imperial Platonism." In Sharples and Sorabji 2007:2.365–377.

Bonazzi, M., and J. Opsomer, eds. 2009. *The Origins of the Platonic System: Platonisms of the Early Empire and Their Philosophical Contexts.* Collection d'Études Classiques 23. Louvain.

Borret, M., ed. 1967–1976. *Contre Celse.* 5 vols. SC 132, 136, 147, 150, 227. Paris.

Bovon, F. 2009. *New Testament and Christian Apocrypha.* Wissenschaftliche Untersuchungen zum Neuen Testament 237. Tübingen.

Brisson, L. 1987. "Amélius: Sa vie, son oeuvre, sa doctrine, son style." *Aufstief und Niedergang der Römischen Welt.* Part 2, *Principat,* vol. 36.2 (ed. W. Haase) 793–860. Berlin.

———. 2002. "Theodorus (19)." *Der Neue Pauly: Enzyklopädie der Antike,* vol. 12.1 (ed. H. Cancik et al.) 328–330. Stuttgart.

Bucur, B. 2009. "The Place of the Hypotyposeis in the Clementine Corpus: An Apology for 'The Other Clement of Alexandria.'" *Journal of Early Christian Studies* 17:313–335.

Burguière, P., and P. Évieux, eds. 1985. *Contre Julien.* Vol. 1. SC 322. Paris.

Burkert, W. 1972. *Lore and Science in Ancient Pythagoreanism.* Trans. E. L. Minar, Jr. Cambridge, MA. (Translation of *Weisheit und Wissenschaft: Studien zu Pythagoras, Philolaos und Platon.* Nuremberg, 1962.)

Burnet, J., ed. 1900–1907. *Platonis opera.* 5 vols. Oxford.

Casey, R. P., ed. and trans. 1934. *The Excerpta ex Theodoto of Clement of Alexandria.* London.

Castellano, A. 1998. *La exégesis de Orígenes y de Heracleón a los testimonios del Bautista.* Santiago, Chile.

Catalogue général des manuscrits des bibliothèques publique des départements. 1849. Vol. 1. Paris.

Chiaradonna, R. 2009. "Autour d'Eudore: Les débuts de l'exégèse des Catégories dans le Moyen Platonisme." In Bonazzi and Opsomer 2009:89–111.

Choufrine, A. 2002. *Gnosis, Theophany, Theosis: Studies in Clement of Alexandria's Appropriation of His Background.* New York.

Chrestou, P. K., ed. 1985. Γρηγορίου τοῦ Παλαμᾶ ἅπαντα τὰ ἔργα. Vol. 10. Ἕλληνες Πατέρες τῆς Ἐκκλησίας 76. Thessalonike.

Cohn, L., and P. Wendland, eds. 1896–1930. *Philonis Alexandrini opera quae supersunt.* 7 vols. in 8. Berlin.

Crouzel, H. 1956. *Théologie de l'image de Dieu chez Origène.* Aubier.

Crum, W. E. 1962. *A Coptic Dictionary.* Oxford.

Curd, P. 2005. *The Legacy of Parmenides: Eleatic Monism and Later Presocratic Thought.* Reprinted, with new introduction. Las Vegas.

Davison, J. E. 1983. "Structural Similarities and Dissimilarities in the Thought of Clement of Alexandria and the Valentinians." *Second Century* 3.4:201–217.

de Falco, V., ed. 1922. *(Iamblichi) theologoumena arithmeticae.* Corrected by U. Klein. Leipzig.

Delatte, A. 1915. *Études sur la littérature pythagoricienne.* Paris.

Derron, P. 1992. "Inventaire des manuscrits des Vers d'or pythagoriciens." *Revue d'histoire des texts* 22:1–17.

Descourtieux, P., ed. 1999. *Stromate VI.* SC 446. Paris.

Desjardins, M. 1986. "The Sources for Valentinian Gnosticism: A Question of Methodology." *Vigiliae Christianae* 40:342–347.

des Places, É., ed. 1974. *Numénius: Fragments.* Paris.

Deuse, W., ed. 1973. *Theodoros von Asine: Sammlung der Testimonien und Kommentar.* Palingenesia 6. Wiesbaden.

Diehl, E., ed. 1971. *Theognis.* Corrected by D. Young. Leipzig.

Diels, H. 1879. *Doxographi graeci.* Berlin.

———. 1882–1895. *Simplicii in Aristotelis physicorum libros octo commentaria.* 2 vols. Commentaria in Aristotelem Graeca 9–10. Berlin.

Diels, H., and W. Kranz, eds. 1957. *Die Fragmente der Vorsokratiker: Griechisch und deutsch.* 8th ed. Hamburg.

Dillon, J. 1987. "Iamblichus of Chalcis (c. 240–325 A.D.)." *Aufstief und Niedergang der Römischen Welt.* Part 2, *Principat,* vol. 36.2 (ed. W. Haase) 862–909. Berlin.

———. 1992. "Pleroma and Noetic Cosmos: A Comparative Study." In Wallis 1992: 99–110.

———. 1996. *The Middle Platonists: 80 BC to AD 220.* Revised edition with a new afterword. Ithaca, NY.

———. 2007. "Numenius: Some Ontological Questions." In Sharples and Sorabji 2007:2.397–402.

Dillon, J., ed., trans., and comm. 1973. *Iamblichi Chalcidensis in Platonis Dialogos Commentariorum Fragmenta*. Leiden.

Dillon, J., and J. Hershbell. 1991. *On the Pythagorean Way of Life: Text, Translation, and Notes*. Texts and Translations 29. Atlanta.

Dodds, E. R. 1928. "The *Parmenides* of Plato and the Origin of the Neoplatonic 'One.'" *Classical Quarterly* 22:129–142.

D'Ooge, M. L., trans. 1926. *Introduction to Arithmetic*. New York.

Dornseiff, F. 1925. *Das Alphabet in Mystik und Magie*. Leipzig.

Dörrie, H., and M. Baltes. 1996. *Der Platonismus in der Antike: Grundlagen, System, Entwicklung*. Vol. 4. Stuttgart-Bad Cannstatt.

Dressaart Lulofs, H. J., ed. 1965. *Aristotelis de generatione animalium*. Oxford.

Drpić, I. 2008. "Art, Hesychasm, and Visual Exegesis: Parisinus Graecus 1242 Revisited." *Dumbarton Oaks Papers* 62:217–247.

Dupont-Sommer, A. 1946. *La doctrine gnostique de la lettre "Wâw," d'après une lamelle araméenne inédite*. Paris.

Düring, I. 1932. *Porphyrios Kommentar zur Harmonielehre des Ptolemaios*. Göteborgs Högskolas Årsskrift 38. Gothenburg.

Düring, I., ed. 1930. *Die Harmonielehre des Klaudios Ptolemaios*. Göteborgs Högskolas Årsskrift 36. Gothenburg.

Edwards, M. J. 1997. "Simon Magus, the Bad Samaritan." *Portraits: Biographical Representation in the Greek and Latin Literature of the Roman Empire* (ed. M. J. Edwards and S. Swain) 69–91. Oxford.

———. 2000. "Clement of Alexandria and his Doctrine of the Logos." *Vigiliae Christianae* 54.2:159–177.

———. 2002. *Origen Against Plato*. Aldershot.

———. 2006. *Culture and Philosophy in the Age of Plotinus*. London.

Estienne, H., ed. 1851. *Thesaurus graecae linguae*. Vol. 8. Paris.

Ferguson, E. 1990. "Was Barnabas a Chiliast? An Example of Hellenistic Number Symbolism in Barnabas and Clement of Alexandria." *Greeks, Romans, and Christians* (ed. D. L. Balch, E. Ferguson, and W. A. Meeks) 157–167. Minneapolis.

Ferwerda, R. 1982. "Plotinus on Sounds: An Interpretation of Plotinus' Enneads V, 5, 5, 19–27." *Dionysius* 6:43–57.

Festa, N., ed. 1891. *Iamblichi de communi mathematica scientia liber*. Corrected by U. Klein. Leipzig.

Festugière, A.-J., ed. and trans. 1966–1968. *Commentaire sur le* Timée. 5 vols. Paris.

Finamore, J. F. 2000. "Iamblichus, the Sethians, and Marsanes." In Turner and Majercik 2000:225–257.

Finamore, J. F., and J. Dillon. 2002. *Iamblichus: De anima. Text, Translation, and Commentary*. Leiden.

Förster, N. 1999. *Marcus Magus: Kult, Lehre und Gemeindeleben einer valentinianischen Gnostikergruppe: Sammlung der Quellen und Kommentar.* Wissenschaftliche Untersuchungen zum Neuen Testament 114. Tübingen.

Fredouille, J.-C., ed. 1980–1981. *Contre les Valentiniens.* SC 280, 281. Paris.

Frickel, J. 1968. *Die Apophasis Megale in Hippolyt's Refutatio (VI 9-18): Eine Paraphrase zur Apophasis Simons.* Orientalia Christiana Analecta 182. Rome.

Friedlein, G., ed. 1873. *Procli Diadochi in primum Euclidis elementorum librum commentarii.* Leipzig.

Funk, W.-P., P.-H. Poirier, and J. D. Turner, eds. 2000. *Marsanès: NH X.* Bibliothèque Copte de Nag Hammadi, Section "Textes" 27. Louvain.

Galtier, M. É. 1902. "Sur les mystères des lettres grecques." *Bulletin de l'Institut français d'archéologie orientale* 2.2:139–162.

Gardthausen, V. 1913. *Griechische Palaeographie.* Vol. 2. Leipzig.

Gersh, S. 1978. *From Iamblichus to Eriugena.* Leiden.

Gevirtz, S. 1969. "Abram's 318." *Israel Exploration Journal* 19.2:110–113.

Giangrande, J., ed. 1956. *Eunapii vitae sophistarum.* Rome.

Gignac, F. 1976. *A Grammar of the Greek Papyri of the Roman and Byzantine Periods.* 2 vols. Milan.

Goodspeed, E. J., ed. 1915. *Die ältesten Apologeten.* Göttingen.

Graef, H. C. 1952. "L'image de Dieu et la structure de l'âme chez les Pères grecs." *Supplément de La vie spirituelle* 22.

Grant, R. M., ed. 1970. *Ad Autolycum.* Oxford.

Gwyn Griffiths, J., ed. 1970. *Plutarch's De Iside et Osiride.* Cambridge.

Haar, S. 2003. *Simon Magus: The First Gnostic?* Beihefte zur Zeitschrift für die neutestamentliche Wissenschaft und die Kunde der älteren Kirche 119. Berlin.

Haase, R. 1969. "Ein Beitrag Platons zur Tetraktys." *Antaios* 11:85–91.

Hamann, C. 1891. *De Psalterio triplici Cusano.* Hamburg.

Hamman, A.-G. 1987. *L'homme image de Dieu: Essai d'une anthropologie chrétienne dans l'Église des cinq premiers siècles.* Paris.

Harding, G. L. 1971. *An Index and Concordance of Pre-Islamic Arabian Names and Inscriptions.* Toronto.

Harmon, A. M., trans. 1915. *Lucian in Eight Volumes.* Vol. 2. Loeb Classical Library 54. Cambridge, MA.

Harms, W. 1975. "Das pythagorische Y auf illustrierten Flugblättern des 17. Jahrhunderts." *Antike und Abendland* 21:97–110.

Harvey, W. W., ed. 1857. *Sancti Irenaei episcopi Lugdunensis libri quinque adversus haereses.* 2 vols. Cambridge.

Hayduck, M., ed. 1891. *Alexandri Aphrodisiensis in Aristotelis metaphysica commentaria.* Commentaria in Aristotelem Graeca 1. Berlin.

Hedrick, C., ed. 1990. *Nag Hammadi Codices XI, XII, XIII.* NHS 28. Leiden.

Heiberg, J. L., ed. 1901. *Sur les dix premiers nombres.* Macon.

Heine, R. E., trans. 1989–1993. *Commentary on the Gospel according to John.* 2 vols. Fathers of the Church 80, 89. Washington.

Heintz, F. 1997. *Simon "le magicien": Actes 8, 5–25 et l'accusation de magie contre les prophètes thaumaturges dans l'antiquité.* Cahiers de la Revue biblique 39. Paris.

Henry, P., and H.-R. Schwyzer. 1951–1982. *Plotini opera.* 3 vols. Oxford.

Henry, R., ed. 1959–1991. *Bibliothèque.* 9 vols. Paris.

Heubner, H., ed. 1983. *Ab excessu divi Augusti.* Stuttgart.

Heyne, C. G. 1834. *Homeri Ilias.* 2 vols. Oxford.

Hilberg, I., ed. 1996. *Sancti Eusebii Hieronymi Epistulae.* 2nd ed. Corpus Scriptorum Ecclesiasticorum Latinorum 54–56. Vienna.

Hill, C. E. 2006. *From the Lost Teaching of Polycarp: Identifying Irenaeus' Apostolic Presbyter and the Author of Ad Diognetum.* Wissenschaftliche Untersuchungen zum Neuen Testament 186. Tübingen.

Hiller, E., ed. 1878. *Theonis Smyrnaei philosophi Platonici expositio rerum mathematicarum ad legendum Platonem utilium.* Leipzig.

Hoche, R., ed. 1866. *Nicomachi Geraseni Pythagorei introductionis arithmeticae libri ii.* Leipzig.

Holl, K., ed. 1922. *Epiphanius.* Vol. 2. *Ancoratus und Panarion.* GCS 31. Leipzig.

Hort, F. J. A., and J. B. Mayor, eds. 1902. *Clement of Alexandria: Miscellanies, Book VII: The Greek Text with Introduction, Translation, Notes, Dissertations, and Indices.* London.

Hubert, C. 1938. *Plutarchi moralia.* Vol. 4. Leipzig.

———. 1954. *Plutarchi moralia.* Vol. 6.1:143–188. Leipzig.

Huffman, C. A. 1993. *Philolaus of Croton: Pythagorean and Presocratic; A Commentary on the Fragments and Testimonia with Interpretive Essays.* Cambridge.

———. 2005. *Archytas of Tarentum: Pythagorean, Philosopher, and Mathematician King.* Cambridge.

Hunger, H., and W. Lackner. 1992. *Katalog der griechischen Handschriften der österreichischen Nationalbibliothek.* Vol. 3, *Codices Theologici 201-337.* Vienna.

Hurtado, L. W. 1998. "The Origin of the Nomina Sacra: A Proposal." *Journal of Biblical Literature* 117:655–673.

Hvalvik, R. 1987. "Barnabas 9.7–9 and the Author's Supposed Use of Gematria." *New Testament Studies* 33:276–282.

Isnardi Parente, M., ed. 1982. *Frammenti.* Naples.

Itter, A. 2009. *Esoteric Teaching in the "Stromateis" of Clement of Alexandria.* Leiden.

Jakab, A. 2001. *Ecclesia alexandrina: Évolution sociale et institutionnelle du christianisme alexandrin (IIe et IIIe siècles).* Christianismes anciens 1. Bern.

Jannaris, A. N. 1907. "The Digamma, Koppa, and Sampi as Numerals in Greek *Classical Quarterly* 1:37–40.

Jaubert, A., ed. 1971. *Épître aux Corinthiens.* SC 167. Paris.

Jordan, D. H. 1913. "Armenische Irenaeusfragmente." *Texte und Untersuchungen* 36.3.

Kahn, C. H. 2001. *Pythagoras and the Pythagoreans: A Brief History.* Indianapolis.

Kalvesmaki, J. 2006. "Formation of the Early Christian Theology of Arithmetic: Number Symbolism in the Late Second and Early Third Century." Ph.D. dissertation, Catholic University of America. Washington.

———. 2007a. "Isopsephic Inscriptions from Iasos (Inscriften von Iasos 419) and Shnān (IGLS 1403)." *Zeitschrift für Papyrologie und Epigraphik* 161:261–268.

———. 2007b. "The Original Sequence of Irenaeus, Against Heresies 1: Another Suggestion." *Journal of Early Christian Studies* 15:407–417.

———. 2008. "Eastern versus Italian Valentinianism?" *Vigiliae Christianae* 62:79–89.

———. Forthcoming. "Pachomius and the Mystery of the Letters." *Asceticism: In Honor of Philip Rousseau* (ed. B. Leyerle and R. Darling Young). South Bend, IN.

Kárpáti, A. 1993. "The Musical Fragments of Philolaus and the Pythagorean Tradition." *Acta Antiqua Academiae Scientiarum Hungaricae* 34:55–67.

Kasser, R., and G. Wurst. 2007. *The Gospel of Judas together with the Letter of Peter to Philip, James, and a Book of Allogenes from Codex Tchacos.* Critical Edition. Washington.

Keil, B. 1894. "Eine Halikarnassische Inschrift." *Hermes* 29:249–280.

Kelly, J. N. D. 1950. *Early Christian Creeds.* London.

Kilburn, K., trans. 1959. *Lucian in Eight Volumes.* Vol. 6. Loeb Classical Library 430. Cambridge, MA.

Kirk, G., J. Raven, and M. Schofield, eds. 1983. *The Presocratic Philosophers.* 2nd ed. Cambridge.

Knorr, W. R. 1993. "Arithmêtikê stoicheiôsis: On Diophantus and Hero of Alexandria." *Historia Mathematica* 20:180–192.

Köhler, F. G., ed. 1974. *Hieroclis in aureum Pythagoreorum carmen commentarius.* Stuttgart.

Kovacs, J. L. 1997. "Concealment and Gnostic Exegesis: Clement of Alexandria's Interpretation of the Tabernacle." *Studia Patristica* 31:414–437.

Kraft, R. A., ed. 1971. *Épître de Barnabé.* SC 172. Paris.

Kroll, W., ed. 1902. *Syriani in metaphysica commentaria.* Commentaria in Aristotelem Graeca 6.1. Berlin.

Kucharski, P. 1952. *Étude sur la doctrine pythagoricienne de la tétrade.* Paris.

Kühn, C. G., ed. 1825. *Claudii Galeni opera omnia.* Vol. 9. Leipzig.

Ladner, G. B. 1953. "The Concept of the Image of God in the Greek Fathers and the Byzantine Iconoclastic Controversy." *Dumbarton Oaks Papers* 7:1–34.

Lamberton, R. 2001. *Plutarch*. New Haven.

Lampe, G. W. H. 1961. *A Patristic Greek Lexicon*. Oxford.

Lampropoulou, S. 1975. "Περὶ τινῶν Πυθαγορείων φιλοσοφικῶν προτύρων παρὰ Πλάτωνι." *Platon* 27:7–19.

Lampros, S. P. 1895. *Κατάλογος τῶν ἐν ταῖς βιβλιοθήκαις τοῦ Ἁγίου Ὄρους ἑλληνικῶν κωδικῶν*. 2 vols. London.

Larfeld, W. 1898. *Handbuch der griechischen Epigraphik*. Vol. 1. Leipzig.

Layton, B., trans. 1987. *The Gnostic Scriptures*. Garden City, NY.

le Boulluec, A. 1982. "Exégèse et polémique antignostique chez Irénée et Clément d'Alexandrie: L'exemple du centon." *Studia Patristica* 17.2:707–713.

Lieberman, S. 1987. "A Mesopotamian Background for the So-called Aggadic 'Measures' of Biblical Hermeneutics?" *Hebrew Union College Annual* 58:157–225.

Long, H. S., ed. 1964. *Vitae philosophorum*. 2 vols. Oxford.

Luz, C. 2010. *Technopaignia: Formspiele in der griechischen Dichtung*. Leiden.

McCarthy, D. P., and A. Breen. 2003. *The Ante-Nicene Christian Pasch: De ratione paschali; The Paschal Tract of Anatolius, Bishop of Laodicea*. Dublin.

Mansfeld, J. 1992. *Heresiography in Context: Hippolytus' Elenchos As a Source for Greek Philosophy*. Leiden.

Marcovich, M., ed. 1986. *Refutatio omnium haeresium*. Patristische Texte und Studien 25. Berlin.

Markschies, C. 1992. *Valentinus Gnosticus? Untersuchungen zur valentinianischen Gnosis. Wissenschaftliche Untersuchungen zum Neuen Testament* 142. Tübingen.

———. 2000. "New research on Ptolemaeus Gnosticus." *Zeitschrift für antikes Christentum* 4.2:225–254.

———. 2003. *Gnosis: An Introduction*. Trans. J. Bowden. London.

Marrou, H.-I., M. Harl, C. Mondésert, and C. Matray, eds. 1960–1970. *Le pedagogue*. 3 vols. SC 70, 108, 158. Paris.

Marx, J. 1905. *Verzeichnis der Handschriften-Sammlung des Hospitals zu Cues bei Bernkastel a./Mosel*. Treves.

Mau, J., and H. Mutschmann, eds. 1914. *Sexti Empirici opera*. 4 vols. Leipzig.

Méautis, G. 1918. *Hermoupolis-la-Grande*. Lausanne.

Mercier, C., ed. 1979. *Quaestiones et solutiones in Genesim: E versione armeniaca*. Les oeuvres de Philon d'Alexandrie 34ab. Paris.

Merki, H. 1952. *Omoiosis Theo: Von der Platonischen Angleichung an Gott zur Gottähnlichkeit bei Gregor von Nyssa*. Paradosis 7. Freiburg.

Miller, E. 1880. "Glossaire grec-latin de la bibliothèque de Laon." *Notices et extraits des manuscrits* 29.2:1–230.

Miziołek, J. 1990. "Transfiguratio Domini in the Apse at Mount Sinai and the Symbolism of Light." *Journal of the Warburg and Courtauld Institutes* 53:42–60.

Mondésert, C., ed. 1954. *Stromate II*. Sources chrétiennes 38. Paris.

Montfaucon, B. de. 1708. *Palæographia græca, sive De ortu et progressu literatum græcarium, et de variis omnium sæculorum scriptionis græcæ generibus: Itemque de abbreviationibus et de notis variarum artium ac disciplinarum*. Paris.

Morani, M., ed. 1987. *Nemesii Emeseni De natura hominis*. Leipzig.

Moreschini, C., ed. 2003. *Commentario al "Timeo" di Platone*. Milan.

Morrow, G. R., and J. M. Dillon, trans. 1987. *Proclus' Commentary on Plato's "Parmenides."* Cambridge.

Mouraviev, S. N. 1984. "Valeurs phoniques et ordre alphabétique en vieux-géorgien." *Zeitschrift der deutschen morgenländischen Gesellschaft* 134:61–83.

Mullach, F. W. A., ed. 1867. *Fragmenta philosophorum Graecorum*. Vol. 2. Paris.

Müller, K. O., ed. 1849. *Fragmenta historicorum graecorum*. Vol. 3. Paris.

Nauck, A. 1886. *Porphyrii philosophi Platonici opuscula selecta*. 2nd ed. Leipzig.

Nestle, Eberhard, Erwin Nestle, and K. Aland, eds. 2004. *Novum Testamentum Graece*. 27th rev. ed. Stuttgart.

Nikulin, D. 2002. *Matter, Imagination, and Geometry: Ontology, Natural Philosophy, and Mathematics in Plotinus, Proclus, and Descartes*. Burlington, VT.

Nock, A. D., ed. 1983. *Corpus Hermeticum*. Trans. A. J. Festugière. 6th ed., 4 vols. Paris.

O'Meara, D. J. 1989. *Pythagoras Revived: Mathematics and Philosophy in Late Antiquity*. New York.

Opsomer, J. 2007. "Plutarch on the One and the Dyad." In Sharples and Sorabji 2007:2.379–395.

Orbe, A. 1976. *Cristología gnóstica: Introducción a la soteriología de los siglos II y III*. Volume 1. Biblioteca de autores cristianos 384. Madrid.

Otto, J. C. T., ed. 1879. *Iustini philosophi et martyris opera: Opera Iustini addubitata*. Vol. 3 of *Corpus apologetarum Christianorum saeculi secundi*. 3rd ed. Jena.

Pack, R. A. 1963. *Artemidori Daldiani Onirocriticon libri v*. Leipzig.

Parrott, D. M., ed. 1991. *Nag Hammadi codices III, 3-4 and V, 1 with Papyrus Berolinensis 8502, 3 and Oxyrhynchus papyrus 1081:* Eugnostos *and* The Sophia of Jesus Christ. NHS 27. Leiden.

Patterson, L. G. 1997. "The Divine Became Human: Irenaean Themes in Clement of Alexandria." *Studia Patristica* 31:497–516.

Pearson, B. A., ed. 1981. *Nag Hammadi Codices IX and X*. NHS 15. Leiden.

Peel, M., ed. 1985. "The Treatise on the Resurrection." *Nag Hammadi Codex I (the Jung Codex)* 22:123–157, 23:137–215. NHS 22–23. Leiden.

Perkins, P. 1992. "Beauty, Number, and Loss of Order in the Gnostic Cosmos." In Wallis 1992:277–296.

Petit, F., ed. 1978. *Quaestiones in Genesim et in Exodum: Fragmenta Graeca*. Les oeuvres de Philon d'Alexandrie 33. Paris.

Pohlenz, M. 1970–1972. *Die Stoa: Geschichte einer geistigen Bewegung*. 4th ed. 2 vols. Göttingen.

Pouderon, B. 2005. "La notice d'Hippolyte sur Simon: Cosmologie, anthropologie et embryologie." *Médecine et théologie chez les Pères de l'Église* (ed. V. Boudon-Millot and B. Pouderon) 49–71. Theologie historique 117. Paris.

Priesendanz, K., and A. Heinrichs, eds. 1973–1974. *Papyri Graecae Magicae: Die Griechischen Zauberpapyri*. 2nd ed., 2 vols. Stuttgart.

Quacquarelli, A. 1973. *L'ogdoade patristica e i suoi riflessi nella liturgia e nei monumenti*. Quaderni di *Vetera Christianorum* 7. Bari.

Quasten, J. 1953. *Patrology*. Vol. 2. *The Ante-Nicene Literature after Irenaeus*. Westminster, MD.

Quispel, G., ed. 1966. *Lettre à Flora*. 2nd ed. SC 24 bis. Paris.

Rackham, H., W. H. S. Jones, and D. E. Eichholz, trans. 1938–1963. *Natural History*. 10 vols. Loeb Classical Library. Cambridge, MA.

Rahlfs, A., ed. 1935. *Septuaginta: Id est Vetus Testamentum graece iuxta LXX interpretes*. Stuttgart.

Ramelli, I. 2011. "Origen's Doctrine of Apokatastasis: A Reassessment." *Origeniana Decima: Origen as Writer* (ed. S. Kaczmarek and H. Pietras) 649–670. Leuven.

Rasimus, T. 2005. "Ophite Gnosticism, Sethianism, and the Nag Hammadi Library." *Vigiliae Christianae* 59:235–263.

Riley, M. T., trans. 1971. "Q. S. Fl. Tertulliani Adversus Valentinianos: Text, Translation, and Commentary." Ph.D. dissertation, Stanford University. http://www.tertullian.org/articles/riley_adv_val/riley_00_index.htm. Accessed October 2005.

Rist, J. M. 1962. "The Neoplatonic One and Plato's Parmenides." *Transactions and Proceedings of the American Philological Association* 93:389–401.

———. 1965. "Monism: Plotinus and Some Predecessors." *Harvard Studies in Classical Philology* 69:329–344.

Robbins, F. E. 1921. "The Tradition of Greek Arithmology." *Classical Philology* 16:97–123.

Robert, L. 1960. "Pas de date 109, mais le chiffre 99, isopséphie de Amen." *Hellenica: Recueil d'épigraphie de numismatique et d'antiquités grecques* 11:310–311. Paris.

Robinson, J. M., ed. 1973. *The Facsimile Edition of the Nag Hammadi Codices*. Vol. 10, *Codices XI, XII, and XIII*. Leiden.

———. 1988. *The Nag Hammadi Library*. Rev. ed. San Francisco.

Rochefort, G., ed. 1963. *[Julien:] Oeuvres completes.* 2 vols. Paris.

Roscher, W. H., ed. 1913. *Die hippokratische Schrift von der Siebenzahl.* Studien zur Geschichte und Kultur des Altertums 6. Paderborn.

Rose, V., ed. 1886. *Aristotelis qui ferebantur librorum fragmenta.* Leipzig.

Ross, W. D., ed. 1924. *Aristotle's Metaphysics.* 2 vols. Oxford.

Rousseau, A., L. Doutreleau, B. Hemmerdinger, and C. Mercier, eds. 1965–1982. *Contre les hérésies.* 10 vols. Sources chrétiennes 263–264, 293–294, 210–211, 100, 152–153. Paris.

Ruelle, C. É., ed. 1899. *Damascii successoris: Dubitationes et solutiones.* Paris.

Runia, D. T. 1995. "Why Does Clement of Alexandria Call Philo 'the Pythagorean'?" *Vigiliae Christianae* 49:1–22.

Safty, E. 2003. *La psyché humaine: Conceptions populaires, religieuses et philosophiques en Grèce, des origines à l'ancien stoïcisme.* Paris.

Sagnard, F-M-M. 1947. *La gnose valentinienne et le témoignage de Saint Irénée.* Études de philosophie médiévale 36. Paris.

Salles-Dabadie, J. M. A. 1969. *Recherches sur Simon le Mage.* Cahiers de la *Revue biblique* 10. Paris.

Salmon, G. 1882. "Monoimus." *Dictionary of Christian Biography, Literature, Sects, and Doctrines* (ed. W. Smith and H. Wace) 3:934–935.

———. 1887. "Ophites." *Dictionary of Christian Biography, Literature, Sects, and Doctrines* (ed. W. Smith and H. Wace) 4:80–88.

Sambursky, S. 1978. "On the Origin and Significance of the Term Gematria." *Journal of Jewish Studies* 29:35–38.

Sandbach, F. H., trans. 1969. *Moralia.* Vol. 15. Loeb Classical Library 429. Cambridge, MA.

Sbordone, F. 1981. "Per la storia antica e recente del numero quattro." *Scritti in onore di Nicola Petruzzellis* 337–342. Naples.

Schenke, H.-M. 1962. *Der Gott "Mensch" in der Gnosis: Ein religionsgeschichtlicher Beitrag zur Diskussion über die paulinische Anschauung von der Kirche als Leib Christi.* Göttingen.

———, ed. 1997. "Das Philippus-Evangelium." *Nag-Hammadi-Codex* II, 3. Berlin.

Schindler, K. 1934. *Die stoische Lehre von den Seelenteilen und Seelenvermögen insbesondere bei Panaitios und Posidonios.* Munich.

Schmidt, C., ed. 1978a. *The Books of Jeu and the Untitled Text in the Bruce Codex.* Trans. V. MacDermot. NHS 13. Leiden.

———. 1978b. *Pistis Sophia.* Trans. V. MacDermot. NHS 9. Leiden.

Schneider, R., ed. 1910. *Grammatici Graeci.* Part 2, vol. 3. *Librorum Apollonii deperditorum fragmenta.* Leipzig.

Schoedel, W. R., ed. 1972. *Legatio and De resurrectione.* Oxford.

Schoedel, W. R., and D. M. Parrott, eds. 1979. "The (First) Apocalypse of James." *Nag Hammadi Codices V, 2-5 and VI with Papyrus Berolinensis 8502, 1 and 4* (ed. D. M. Parrott) 65–103. NHS 11. Leiden.

Sharples, R., and R. Sorabji, eds. 2007. *Greek and Roman Philosophy 100 BC-200 AD*. 2 vols. BISC Supplement 94. London.

Shaw, G. 1999. "Eros and Arithmos: Pythagorean Theurgy in Iamblichus and Plotinus." *Ancient Philosophy* 19:121–143.

Sieveking, W. 1929. *Plutarchi moralia*. Vol. 3. Leipzig.

Slaveva-Griffin, S. 2009. *Plotinus on Number*. Oxford.

Spanneut, M. 1957. *Le Stoïcisme des pères de l'église*. Paris.

Staab, G. 2009. "Das Kennzeichen des neuen Pythagoreismus innerhalb der kaiserzeitlichen Platoninterpretation: 'Pythagoreischer' Dualismus und Einprinzipienlehre im Einklang." In Bonazzi and Opsomer 2009:55–88.

Stähle, K., ed. 1931. *Die Zahlenmystik bei Philon von Alexandreia*. Leipzig.

Stählin, O., ed. 1909. *Clemens Alexandrinus*. Vol. 3, *Stromata Buch VII und VIII*. GCS 17. Berlin. Second ed. 1970 by L. Früchtel.

Stead, G. C. 1969. "The Valentinian Myth of Sophia." *Journal of Theological Studies* 20:75–104.

———. 1980. "In Search of Valentinus." *The Rediscovery of Gnosticism*. Vol. 1, *The School of Valentinus* (ed. B. Layton) 75–95. Leiden.

Stein, L. 1886–1888. *Die Psychologie der Stoa*. 2 vols. Berlin.

Stevenson, K. 2001. "Animal Rites: The Four Living Creatures in Patristic Exegesis and Liturgy." *Studia Patristica* 34:470–492.

Tarán, L. 1981. *Speusippus of Athens: A Critical Study with a Collection of the Related Texts and Commentary*. Leiden.

———, ed. 1969. "Asclepius of Tralles: Commentary to Nicomachus' Introduction to Arithmetic." *Transactions of the American Philosophical Society*, n.s. 59.4: 24–72.

Taylor, T., trans. 1820. *The Commentaries of Proclus on the Timaeus of Plato, in Five Books: Containing a Treasury of Pythagoric and Platonic Physiology*. London.

Teodorsson, S.-T. 1989–1996. *A Commentary on Plutarch's Table Talks*. 3 vols. Studia Graeca et Latina Gothoburgensia 51, 53, 62. Gothenburg.

Ter Měkĕrttschian, K., and S. G. Wilson, eds. 1919. "The Proof of the Apostolic Preaching with Seven Fragments." *Patrologia Orientalis* 12:655–744.

Terian, A. 1984. "A Philonic Fragment on the Decad." *Nourished with Peace: Studies in Hellenistic Judaism in Memory of Samuel Sandmel* (ed. F. E. Greenspahn, E. Hilgert, and B. L. Mack). Chico, Calif.

Thesleff, H. 1961. *An Introduction to the Pythagorean Writings of the Hellenistic Period*. Turku.

———, ed. 1965. *The Pythagorean Texts of the Hellenistic Period*. Turku.

Thomassen, E. 1994–1995. "L'histoire du valentinisme et le Traité Tripartite." *Annuaire Ecole pratique des hautes études, Section sciences religieuses* 103: 301–303.

———. 1995. "Notes pour la délimitation d'un corpus valentinien à Nag Hammadi." *Les textes de Nag Hammadi et le problème de leur classification: Actes du colloque tenu à Québec du 15 au 19 septembre 1993* (ed. L. Painchaud and A. Pasquier) 243–259. Bibliothèque copte de Nag Hammadi. Section "Textes" 3. Quebec.

———. 2000. "The Derivation of Matter in Monistic Gnosticism." In Turner and Majercik 2000:1–18.

———. 2006. *The Spiritual Seed: The Church of the "Valentinians."* NHS 60. Leiden.

Thomassen, E., ed. 1989. *Le traité tripartite (NH I, 5).* BCNH.T 19. Quebec.

Tod, M. N. 1979. *Ancient Greek Numerical Systems.* Chicago.

Tornau, C. 2001. "Die Prinzipienlehre des Moderatos von Gades: Zu Simplikios in Phys. 230, 34–231, 24 Diels." *Rheinisches Museum für Philologie* 143:197–220.

Trapp, M. 2007. "Neopythagoreans." In Sharples and Sorabji 2007:2.347–363.

Turner, J. D. 2006. "The Gnostic Sethians and Middle Platonism: Interpretations of the 'Timaeus' and 'Parmenides.'" *Vigiliae Christianae* 60:9–64.

———, ed. 1990. "NHC XI, 2: A Valentinian Exposition 22, 1–39, 39." *Nag Hammadi Codices XI, XII, XIII* (ed. C. W. Hedrick) 89–172. NHS 28. Leiden. (Introduction by E. Pagels; notes by Turner and Pagels.)

Turner, J. D., and R. Majercik, eds. 2000. *Gnosticism and Later Platonism: Themes, Figures, and Texts.* SBL Symposium 12. Atlanta.

Uhlig, G., and A. Hilgard, eds. 1883–1901. *Grammatici Graeci.* Part 1, 3 vols. Leipzig.

Unger, D. J., and J. J. Dillon, trans. 1992. *St. Irenaeus of Lyons Against the Heresies.* Ancient Christian Writers 55. New York.

van den Hoek, A., ed. 2001. *Stromate IV.* Trans. C. Mondésert. SC 463. Paris.

van Straaten, M. 1946. *Panétius, sa vie, ses écrits et sa doctrine.* Amsterdam.

van Weddingen, R.-E., ed. 1957. *Disputatio de Somnio Scipionis.* Collection Latomus 27. Brussels.

Vidman, L. 1975. "Koppa Theta = Amen in Athen." *Zeitschrift für Papyrologie und Epigraphik* 16:215–216.

Wachsmuth, C., and O. Hense, eds. 1884–1912. *Ioannis Stobaei Anthologium.* 5 vols. Berlin.

Waldstein, M., and F. Wisse, eds. 1995. *The Apocryphon of John: Synopsis of Nag Hammadi Codices II, 1, III, 1, and IV, 1 with BG 8502, 2.* Leiden.

Wallis, R. T., ed. 1992. *Neoplatonism and Gnosticism.* Albany.

Waszink, J. H., ed. 1947. *De anima: Edited with Introduction and Commentary.* Amsterdam.

Waterfield, R., trans. 1989. *The Theology of Arithmetic: On the Mystical, Mathematical and Cosmological Symbolism of the First Ten Numbers*. Grand Rapids, MI.

Wendland, P., ed. 1916. *Refutatio omnium haeresium*. GCS 26. Leipzig.

Whittaker, M., ed. 1967. *Die apostolischen Väter*. Vol. 1, *Der Hirt des Hermas*. GCS 48. 2nd ed. Berlin.

Williams, B. P., and R. S. Williams. 1995. "Finger Numbers in the Greco-Roman World and the Early Middle Ages." *Isis* 86:587–608.

Williams, F., trans. 1987. *The Panarion of Epiphanius of Salamis*. NHS 35–36. Leiden.

Williams, M. A. 1985. *The Immovable Race: A Gnostic Designation and the Theme of Stability in Late Antiquity*. Leiden.

———. 1996. *Rethinking "Gnosticism": An Argument for Dismantling a Dubious Category*. Princeton.

Wucherpfennig, Ansgar. 2002. *Heracleon Philologus: Gnostische Johannesexegese im zweiten Jahrhundert*. WUNT 142. Tübingen.

Wünsch, R. 1898. *Liber de mensibus*. Leipzig.

Index

156, 179, 181; intellectual, 179; monadic, 157; nonalphabetic (παράσημα), 67–68, 144; perfect (*see* perfect numbers); substantial, 157; versus numerable things, 96, 178–179

Numenius of Apamea, 8, 154–155, 159, 166

numeration, 122; Georgian alphabetic, 145; Greek alphabetic, 67, 76, 77, 122, 143–146; Hebrew alphabetic, 122

numeri ex regula, 115–116, 121

numerology: defined, 4–5; Marcus, 82–83

odd numbers, 9–10, 11, 13; male, 50

ogdoad: Clement of Alexandria, 137–138, 148–149; criticism by Irenaeus, 111, 112, 114; episemos, 71, 72; evident in material world, 52; Hebdomad, 137, 140, 149; in human body, 52; Marcus, 69, 74–75, 76, 77, 78, 79–80; proportions, 50; Valentinian, 33–34, 50, 54–55, 104; proof texts, 45–46, 111; Wisdom, 42. *See also* 8

omicron. *See* o (*Greek letters at end of index*)

On the Decalogue, 128

one (ἕν), 32–33; aeon, 32, 74; and indefinite dyad, 16, 18–19, 95, 156; Church, 127; Clement of Alexandria, 126–127; described by Plotinus, 156; immutability, 180; term avoided by Irenaeus, 116; testament, 127; transcended by God, 127; versus monad, 87, 96, 156, 175–182. *See also* 1; Monad

one and the many, 15–16, 60

one Monad (μία μονάς), 86, 87, 101

Ophites, 85

Origen, 172

Orphic mythology, 15

orthodoxy: defined, 3–4; number symbolism, 114–116, 150–151, 171–173; varying expressions, 126; versus heresy, 3–4

orthography. *See* spelling variations

Osiris: number symbolism, 13

Palamedes, 107

Pantaenus, 126

parables, 30, 78–79, 80, 113

parakuisma (παρακύϊσμα), 144

Paraphrase of the "Apophasis Megale," 94–102; compared to Clement of Alexandria, 150; compared to Monoïmus, 101; compared to Valentinians, 101–102

parasema (παράσημα), 144

Parmenides, 15

Pascha: number symbolism, 90–91

Passion (aeon), 41, 78

Pentateuch: derivation, 90; number symbolism, 92, 100–101, 122

perfect numbers, 9, 31, 53–54, 31, 53–54, 119, 150–151, 167–168

Pherecydes, 15

Philo: numerical terminology, 181

Philo of Alexandria, 8–10, 14; compared to Valentinians, 60; powers, 89; used by Clement of Alexandria, 129, 130

Philolaus, 50, 179, 184; lost work *On Nature*, 184

plagues: ten, 90, 91

Plato: and Monoïmus, 90, 93; cave allegory, 21; compared to Moses, 90; dualism, 15–16; number symbolism, 11; philosophy of numbers, 16, 49, 179, 180;

CPSIA information can be obtained
at www.ICGtesting.com
Printed in the USA
LVHW030747020223
738080LV00002B/8